Springer Theses

Recognizing Outstanding Ph.D. Research

More information about this series at http://www.springer.com/series/8790

Aims and Scope

The series "Springer Theses" brings together a selection of the very best Ph.D. theses from around the world and across the physical sciences. Nominated and endorsed by two recognized specialists, each published volume has been selected for its scientific excellence and the high impact of its contents for the pertinent field of research. For greater accessibility to non-specialists, the published versions include an extended introduction, as well as a foreword by the student's supervisor explaining the special relevance of the work for the field. As a whole, the series will provide a valuable resource both for newcomers to the research fields described, and for other scientists seeking detailed background information on special questions. Finally, it provides an accredited documentation of the valuable contributions made by today's younger generation of scientists.

Theses are accepted into the series by invited nomination only and must fulfill all of the following criteria

- They must be written in good English.
- The topic should fall within the confines of Chemistry, Physics, Earth Sciences, Engineering and related interdisciplinary fields such as Materials, Nanoscience, Chemical Engineering, Complex Systems and Biophysics.
- The work reported in the thesis must represent a significant scientific advance.
- If the thesis includes previously published material, permission to reproduce this must be gained from the respective copyright holder.
- They must have been examined and passed during the 12 months prior to nomination.
- Each thesis should include a foreword by the supervisor outlining the significance of its content.
- The theses should have a clearly defined structure including an introduction accessible to scientists not expert in that particular field.

Yangyang Cheng

Search for Dark Matter Produced in Association with a Higgs Boson Decaying to Two Bottom Quarks at ATLAS

Doctoral Thesis accepted by The University of Chicago, Chicago, Illinois, USA

Yangyang Cheng
Cornell Laboratory for Accelerator-based
 ScienceS and Education (CLASSE)
Cornell University
Ithaca, NY, USA

ISSN 2190-5053   ISSN 2190-5061   (electronic)
Springer Theses
ISBN 978-3-319-44217-4   ISBN 978-3-319-44218-1   (eBook)
DOI 10.1007/978-3-319-44218-1

Library of Congress Control Number: 2016949554

© Springer International Publishing Switzerland 2017
This work is subject to copyright. All rights are reserved by the Publisher, whether the whole or part of the material is concerned, specifically the rights of translation, reprinting, reuse of illustrations, recitation, broadcasting, reproduction on microfilms or in any other physical way, and transmission or information storage and retrieval, electronic adaptation, computer software, or by similar or dissimilar methodology now known or hereafter developed.
The use of general descriptive names, registered names, trademarks, service marks, etc. in this publication does not imply, even in the absence of a specific statement, that such names are exempt from the relevant protective laws and regulations and therefore free for general use.
The publisher, the authors and the editors are safe to assume that the advice and information in this book are believed to be true and accurate at the date of publication. Neither the publisher nor the authors or the editors give a warranty, express or implied, with respect to the material contained herein or for any errors or omissions that may have been made.

Printed on acid-free paper

This Springer imprint is published by Springer Nature
The registered company is Springer International Publishing AG
The registered company address is Gewerbestrasse 11, 6330 Cham, Switzerland

Supervisor's Foreword

I am very pleased to introduce this work by my former student, Dr. Yangyang Cheng, who developed a novel method to search for dark matter. Dark matter constitutes 85 % of the matter in the universe, and it played a major role in the development of the universe we see today. However, we don't know what dark matter is. The most likely candidate, a weakly interacting massive particle, could be produced in the high-energy proton–proton collisions at CERN's Large Hadron Collider (LHC). This analysis is unique in looking for dark matter produced together with a Higgs boson that decays into its dominant decay mode, a pair of b quarks. If dark matter were seen in this mode, we would learn directly about the production mechanism because of the presence of the Higgs boson. This thesis develops the search technique and presents the most stringent production limit to date. It is a great honor for Yangyang that her thesis is being published by Springer.

Kersten Distinguished Service Professor of Physics　　　　　　　　Melvyn Shochet
University of Chicago

Abstract

This thesis presents a search for dark matter production in association with a Higgs boson decaying to a pair of bottom quarks, using data from 20.3 fb^{-1} of proton–proton collisions at a center-of-mass energy of 8 TeV collected by the ATLAS detector at the LHC. The dark matter particles are assumed to be weakly interacting massive particles and can be produced in pairs at collider experiments. Events with large E_T^{miss} are selected when produced in association with high momentum jets, of which at least two are identified as jets containing b-quarks consistent with those from a Higgs boson decay. To maintain good detector acceptance and selection efficiency of the signal across a wide kinematic range, two methods of Higgs boson reconstruction are used. The Higgs boson is reconstructed either as a pair of small-radius jets both containing b-quarks, called the "resolved" analysis, or as a single large-radius jet with substructure consistent with a high momentum $b\bar{b}$ system, called the "boosted" analysis. The resolved analysis is the focus of this thesis. The observed data are found to be consistent with the expected Standard Model backgrounds. The result from the resolved analysis is interpreted using a simplified model with a Z' gauge boson decaying into different Higgs bosons predicted in a two-Higgs-doublet model, of which the heavy pseudoscalar Higgs decays into a pair of dark matter particles. Exclusion limits are set in regions of parameter space for this model. Model-independent upper limits are also placed on the visible cross sections for events with a Higgs boson decaying into $b\bar{b}$ and large missing transverse momentum with thresholds ranging from 150 to 400 GeV.

Acknowledgments

Y^2: How many pages of acknowledgments am I allowed to write?
Mel: Keep it to one.

As a surprise to no one but probably a disappointment to many, this acknowledgment ended up being longer than hoped for, but too short to contain every thought and every emotion, every progress and every notable moment, every person that made a difference in my life, and the tribute each one of them deserve.

There is no space big enough for that.

First of all, my heartfelt gratitude to my advisor, Mel Shochet. I still recall vividly the very first time I saw Mel in person, early July of 2009, ATLAS week. As a 19-year-old who practically walked out of her college graduation ceremony with luggage in hand onto a flight from Shanghai to Geneva, I got my first taste of CERN. One of Chicago's postdocs at the time, Monica Dunford, gave a talk. Toward the end of her presentation, a question was raised, and there from the front-left bow of the room, a distinguished-looking gentleman in a crisp white shirt shot up and offered a concise yet forceful defense. "Is that Professor Shochet?" I asked Imai, then a grad student at Chicago, who was sitting next to me and my "guide" for the day. The answer was in the affirmative. I gazed from the back of the auditorium in awe, not just for the name that basically rings HEP legend but also for it was the first time I got a glimpse of Mel's fierce intellect, the sharpness of a diamond cutter. Over six full years have passed, during which time I had one of the greatest fortunes of my life: I became Mel's student. Through these six-plus years, I continue to find myself in awe in Mel's presence. I cannot imagine a better advisor or role model, within and beyond the realm of science. If there is anything that matches the brilliance of his mind, it is the integrity of Mel's character. Mel has taught me how to be a good physicist and inspires me to be a better person.

And many thanks to Bjoern "Dr. Bear" Penning. The one who led me to the "dark" side of the universe. My closest colleague, before he moved on to the other side of the Atlantic and the LHC ring. My dear friend and mentor. The one whom I'd call at 3 a.m. in moments of deep frustration or sorrow. The one who bears (pun intended) both my neurotic tears and even more ridiculous jokes. The one who has offered me boundless advice, support, and the most generous encouragements.

Before he departed Chicago for London leaving me with the daunting task of a full analysis by myself in the early fall of 2014, Bjoern stood by my office door, "You can do this. You got it in here to make it happen." He looked me in the eye and pointed his fist to his heart.

To Young-Kee. A force of nature and an inspiration to anyone lucky enough to be in the vicinity of what Ho Ling used to call "Her Majesty." When you've gotten to know her as an explorer of the subatomic world, a leader of the unruly academic minds, and an organizer of simply the best parties and wonder how one could possibly do it all, wait till you've seen her dance.

To Bill Murray, the physicist. I have often wondered what I've done to deserve such great fortune in my career, the amazing people that I happen to cross or share paths with, the consequence of such encounters being the gift that keeps on giving. Bill is such an example. As the Editorial Board chair of my thesis analysis assigned by some magic in ATLAS management, Bill has, in the words of Mel, "gone above and beyond the call of duty," both with his stewardship of the analysis, the intellectual rigor he injected into the process, and the incredible support he continues to offer me well beyond the scope and duration of this analysis paper. He called me an "irrepressible spirit," something I'd often remind myself of and try to live up to. He has a great sense of humor too. And I heard he is a wonderful dancer. My favorite Bill Murray by far, and that is saying a lot.

I realize at this point there is no justifiable order to keep in writing to the people I'm immensely thankful for. See the following paragraphs as the many parallel dimensions unfolded from the rich life in Chicago I get to call my own because of you. I love you all.

To Jordan. We lived across the street from each other since our first year. We TAed the same courses. We worked on the same experiment. We worked on multiple shared projects for the same experiment. Your clarity in physics. Your work ethic. Your kind heart and gentle soul. I see boundless potential within you and am only so lucky to call you my bestie.

To Joakim. You often say how much I've been a part of your Chicago experience. What I have not told you enough is how much you've been a part of mine. Whatever name we give each other is beside the point. My life is forever different because of you and for the much better.

To Mike, Steven, and Sophie, for their invaluable help through the nitty-gritties of the analysis. To John Alison, for holding the shield and the whip. Please don't generate a mental image of that. Or better, please do.

To Sam, Ho Ling, Jake, Mik, Travis, Radha, and the rest of the gang. I remember sitting in the classroom in KPTC for the first time during orientation week, thinking what kind of mistake it must have been to place me in the same incoming class with all of you, smarter, better trained, and harder working than I ever imagined myself to be. I tried to keep up with you. I gradually realized I was one of you. I found my identity in your company. Thank you for the friendship and the constant, unconditional source of support.

To Guido, Joe "Dr. Tug-Tug" Tuggle, Antonio, Eric, Peter, Jamie, Max, Monica, Lauren, and Reina. The postdocs, past and present. For everything you've taught me, both in physics and in life, both directly and by example. You are all so amazing; I decided I want to be one of you, a postdoc.

To Anton, the senior grad student of Mel when I joined the group, who set the bar high. I hope I did not lower it too much, at least besides our earning power postgraduation. Please consider giving back to the department; an endowed professorship sounds nice.

To Patrick, Karol, and Will. The young ones. You guys are in one of the best places to be students of particle physics and rightfully so. Please help restore sanity and normalcy to this amazing institution when it gets rid of me.

To Tongyan and LianTao. For conceiving the elegant dark matter models I spent a good portion of my graduate school searching for. For all the patience in explaining them in detail. For showing me how much fun and fruitful it can be to have theory collaborators at corridor's length. LianTao served on my thesis committee. I did not find the signals his models predicted. The superb theorist as he is, he did not blame me the experimentalist, but probably secretly blames nature.

To David, Beth, and Natalie. Three generations of neutrinos. Three awesome Schmitz.

To Zheng-Tian. When my age barely grazed double digits, I remember walking through the campus I grew up in on steamy summer nights, staring into news clippings behind glass casings of some promising young physicist winning some esteemed position at some place with a fancy name like Argonne. The young physicist had a pretty cool name too, "explore and conquer the universe." Years later I would enter the same college program that continues to list you as one of its most prized alumni and continue on this shared path across the oceans to this campus. Thank you for being one of my childhood science heroes. And thank you for allowing me the privilege to have you on my thesis committee.

To the Institute of Politics, Steve, Darren, Christine, the rest of the staff, the students, and the guests that I've had the fortune to meet, many of whom I've gotten to know well or even worked closely with. Thank you for welcoming me, a Chinese-born physicist bit by the political bug, with open arms; for showering me with some of the most incredible opportunities beyond my wildest imagination; for teaching me about leadership, communications, and organization; and for trusting me with institutional responsibilities to practice them. Thank you for being a harbor for my other passions in life, for being the balance during the most hectic times, and for making me a more focused and better physicist in the end, however unexpectedly, but it all makes sense. To David Axelrod, one of the most brilliant political minds of a generation with the biggest heart, for showing me what it is like to have conviction beyond the confines of individual existence, for letting me continue to believe one could fight deep and dirty in the trenches, but still gaze at the stars.

To my father, my hero, my champion, my Northern Star. To my mother. To my family, my bloodline, my heritage. For everything we've been through together that made me who I am, the memories of which continue to serve as my greatest source of strength. Certain things are simply dribbles in sand; love is carved in stone. To the little girl scrappy and stubborn in the face of everything life threw her way, thank you for hanging in there: if only I could go back in time some 15 years and let you know, that even in your worst nightmares, your dreams are valid.

To the city of Chicago. As I would often put it, the moment my plane touched down at O'Hare some six-ish years ago, Chicago became home, the first place I've ever lived that offered me a sense of belonging and gave me the most fulfilling life within my reach. The answer always comes so easily whenever I'm asked where I'm from, "Chicago." "You can take the girl out of Chicago. You can never take the Chicago out of the girl."

To everyone, the ones mentioned, the many more left unsaid. The words "thank you" could not be repeated enough. For the differences you've made in my life, I could only try to honor that by aiming to be the best in the many walks of life you've set examples in and striving to be, to others, what you've meant to me.

This work is dedicated to all of you.

Contents

1 **Introduction** 1
  References 3
2 **The Standard Model and Beyond** 5
  2.1 The Standard Model 5
      2.1.1 Bosons 6
      2.1.2 Fermions 7
      2.1.3 Gauge Theory 7
  2.2 Spontaneous Symmetry Breaking and the Higgs Boson 8
  2.3 Physics Beyond the Standard Model 9
  References 11
3 **An Overview of Dark Matter** 13
  3.1 Standard Cosmology 13
  3.2 Relic Density 16
  3.3 Evidence of Dark Matter 17
  3.4 Theories of Dark Matter Candidates 18
  3.5 Experimental Methods 19
      3.5.1 Direct Detection 20
      3.5.2 Indirect Detection 21
      3.5.3 Collider Production 21
  References 23
4 **The ATLAS Experiment at the Large Hadron Collider** 25
  4.1 The Large Hadron Collider 25
  4.2 The ATLAS Detector 27
      4.2.1 Inner Detector 31
      4.2.2 Calorimeters 33
      4.2.3 Muon Spectrometers 34

	4.3	Trigger and Data Acquisition System	35
		4.3.1 The ATLAS Trigger System	36
		4.3.2 Fast Tracker Track Trigger Upgrade	37
	References		38
5	**Dark Matter Searches at ATLAS**		**41**
	5.1	Theoretical Models	41
		5.1.1 Effective Field Theory Framework	42
		5.1.2 Simplified Models	44
	5.2	"Mono-jet" and other "Mono-X" Searches	45
	5.3	Dark Matter Search with Heavy Flavor	46
	References		47
6	**Dark Matter + Higgs($\to b\bar{b}$): Overview**		**49**
	6.1	Physics Motivation	49
	6.2	Signal Models	50
		6.2.1 EFT Models	50
		6.2.2 Simplified Models	51
	6.3	Analysis Channels	51
		6.3.1 Resolved Analysis	51
		6.3.2 Boosted Analysis	51
	References		52
7	**Dark Matter + Higgs($\to b\bar{b}$): Z'-2HDM Simplified Model**		**55**
	7.1	Introduction	55
		7.1.1 Z'-2HDM Model	55
		7.1.2 Z' Constraints	57
		7.1.3 "Mono-Higgs" Signal	59
		7.1.4 Dark Matter Coupling to the Higgs Sector	59
	7.2	Parameter Space and Kinematic Dependencies	61
	7.3	Simulated Signal Samples	63
	References		63
8	**Dark Matter + Higgs($\to b\bar{b}$): Physics Objects**		**67**
	8.1	Data and Simulated Background Processes	67
		8.1.1 Data Sample	67
		8.1.2 Simulated Background Samples	68
	8.2	Trigger	70
		8.2.1 E_T^miss Trigger	70
		8.2.2 Other Triggers	77
	8.3	Final-State Observables: Definition and Selection	77
	References		81
9	**Dark Matter + Higgs($\to b\bar{b}$): Event Selection**		**85**
	9.1	Event Preselection	85
		9.1.1 BCH Cleaning	86
		9.1.2 Data Quality	86

		9.1.3	Vertex Selection	86
		9.1.4	Trigger	87
		9.1.5	Event Cleaning	87
		9.1.6	Jet Cleaning	87
		9.1.7	Jet Vertex Fraction	88
		9.1.8	Jet Multiplicity	88
		9.1.9	E_T^{miss}	88
		9.1.10	Lepton Veto	88
	9.2	Selection of $E_T^{miss} + h(\to b\bar{b})$ Signal		88
		9.2.1	Signal Selection	89
		9.2.2	Signal Selection Efficiency	90
	References			97
10	**Dark Matter + Higgs($\to b\bar{b}$): Background Processes**			99
	10.1	Background Overview		99
	10.2	Simulated Background Processes		100
		10.2.1	W + Jets Control Region	101
		10.2.2	Top Quark Control Region	103
		10.2.3	1-Lepton Validation Region	105
	10.3	Data-Driven Background Processes		107
		10.3.1	Estimation of Multijet Background	108
		10.3.2	Estimation of $Z(\to \nu\bar{\nu})$ + jets Background	116
	10.4	0-Lepton Validation Region		131
	References			136
11	**Dark Matter + Higgs($\to b\bar{b}$): Systematic Uncertainties**			139
	11.1	Sources of Systematic Uncertainties		139
	11.2	Signal Theoretical Uncertainties		143
	11.3	$Z(\to \nu\bar{\nu})$ + Jets Background Systematic Uncertainties		146
	11.4	Systematic Uncertainties in Signal Region		149
	References			154
12	**Dark Matter + Higgs($\to b\bar{b}$): Results**			157
	12.1	Dark Matter + Higgs($\to b\bar{b}$) Signal Region		157
		12.1.1	Event Yield	158
		12.1.2	Final Selections for $DM + h(b\bar{b})$ by $m_{Z'}$ and m_A	161
		12.1.3	Kinematic Distributions	162
	12.2	Statistical Interpretation		163
	12.3	Constraints on Z'-2HDM Model		166
		12.3.1	$m_{Z'}$–tan β Plane	169
	12.4	Model-Independent Upper Limit		172
	References			173
13	**Conclusion**			175

Chapter 1
Introduction

The existence of dark matter (DM) is one of the most striking evidences of physics beyond the Standard Model. Based on standard cosmology theory and observations, the total mass–energy of the known universe contains 4.9 % visible matter, 26.8 % DM, and 68.3 % dark energy [1]; in other words, of all the matter in our universe, there are about five times more DM compared with standard, visible matter. Despite its abundance and a plethora of direct and indirect evidence, the properties of DM, its physical laws, and how it connects to Standard Model (SM) particles remain unknown.

There are mainly three classes of experiments to search for DM: direct detection experiments that measure DM particles from the cosmos recoiling off nucleons of ordinary matter, indirect detection experiments that search for deviations in known particle distributions in the sky which may be a result of dark matter annihilation, and collider experiments where dark matter may be produced and detected as a form of missing energy. The three methods are complementary in determining if a signal observed is indeed from DM and the nature of DM [2].

The leading hypothesis suggests that most of the DM is in the form of stable, electrically neutral, massive particles, i.e., Weakly Interacting Massive Particles [3]. This scenario gives rise to a potential signature at a proton–proton collider where one or more Standard Model (SM) particles, "X," are produced and detected, recoiling against missing transverse momentum (with magnitude E_T^{miss}) associated with the noninteracting DM. Two approaches are commonly used to model generic processes yielding a final state with a particle X recoiling against a system of noninteracting particles. One option is to use nonrenormalizable operators in an effective field theory (EFT) framework [4], where particles that mediate the interactions between DM and SM particles are too heavy to be produced directly in the experiment and are described by contact operators. Alternatively, simplified models that are characterized by a minimal number of renormalizable interactions and hence explicitly include the particles at higher masses can be used [5]. The EFT

© Springer International Publishing Switzerland 2017
Y. Cheng, *Search for Dark Matter Produced in Association with a Higgs Boson Decaying to Two Bottom Quarks at ATLAS*, Springer Theses,
DOI 10.1007/978-3-319-44218-1_1

approach is more model-independent, but is not valid when a typical momentum transfer of the process approaches the energy scale of the contact operators that describe the interaction. Simplified models do not suffer from this concern, but include more assumptions by design and are therefore less generic. The two approaches are thus complementary and both are used in collider searches for DM.

Recent searches at the Large Hadron Collider (LHC) consider "X" to be a hadronic jet [6, 7], heavy-flavor jet [8, 9], photon [10, 11], or W/Z boson [12, 13]. The discovery of the Higgs boson h [14, 15] provides a new opportunity to search for DM production via the $h + E_T^{miss}$ signature [16–18]. In contrast to most of the aforementioned probes, the visible Higgs boson is unlikely to have been radiated from an initial-state quark or gluon, and the signal would give insight into the structure of DM coupling to SM particles. The observed final states are E_T^{miss} plus the Higgs decay products, with an invariant mass constrained to be relatively close to the Higgs mass of around 125 GeV.

The thesis describes a search for WIMP pair production in association with a Higgs boson using the data set corresponding to $20.3\,\text{fb}^{-1}$ of pp collisions at a center-of-mass energy of 8 TeV collected by the ATLAS detector at the Large Hadron Collider during the year 2012. Specifically, we look at the channel where the Higgs boson decays to a pair of bottom quarks, taking advantage of the large branching ratio of the $b\bar{b}$ mode. Two methods of Higgs boson reconstruction are used, the "resolved" channel where the Higgs boson is reconstructed as two separate b-quark jets, and the other "boosted" channel where the Higgs boson is reconstructed as a single large-radius jet using jet substructure techniques. Both EFT models and a simplified model of DM production with a Higgs boson are considered. After studying the interplay between the two sets of models and analysis channels, there is a clear advantage of one analysis channel over the other for either set of models based on signal sensitivity, and for simplicity, the results for the simplified model are given using the resolved analysis, and the EFT models are interpreted using the boosted analysis.

The focus of this thesis is on the resolved channel of the analysis, completed by the author. The results are interpreted as constraints in regions of parameter space for a simplified model with a Z' gauge boson and two Higgs doublets (Z'-2HDM), where the DM is coupled to the heavy pseudoscalar Higgs A^0, i.e., $Z' \to hA^0 \to bb\chi\chi$. Model-independent upper limits on the visible cross-section for final states with large missing energy and a Higgs boson decaying to two b's are also given.

This thesis is organized as follows. Chapter 2 describes the Standard Model (SM) of particle physics, including the Higgs boson and its discovery, and a sketch of potential physics beyond-the-Standard-Model (BSM) that are relevant to the content of this thesis. Chapter 3 gives an overview of dark matter, starting from a brief description of standard cosmology (Sect. 3.1), to introducing the most important cosmological evidence for the existence of dark matter (Sect. 3.3) and the leading candidates (Sect. 3.4), and finally Sect. 3.5 describes the three complementary methods of dark matter searches at experiments, one of which being collider production, which brings the reader to Chap. 4 on the Large Hadron Collider, the ATLAS detector, and its trigger system for data acquisition. Chapter 5 gives a brief overview of important theoretical models and corresponding existing search channels for DM production at ATLAS, and Chap. 6 brings the focus to the search for DM produced in association with a Higgs boson decaying to two b-quarks, after which the analysis in the resolved channel of the $E_\mathrm{T}^\mathrm{miss} + h(\to b\bar{b})$ final state is given in detail. Chapter 7 describes the Z'-2HDM simplified model. Chapter 8 defines the physics objects and triggers used in this analysis, and their reconstruction and selection methods. Chapter 9 gives the selection criteria used to select the $E_\mathrm{T}^\mathrm{miss} + h(\to b\bar{b})$ signal, and the corresponding detector acceptance and selection efficiency for the Z'-2HDM signal. The main background processes in this analysis, how they are modeled and how well the background estimation describes data in various control regions and validation regions, are described in detail in Chap. 10. The systematic uncertainties associated with both simulated background and signal processes are evaluated in Chap. 11. Finally, Chap. 12 gives the results of this analysis, both in terms of constraints in regions of parameter space for the Z'-2HDM simplified model, and model-independent upper limits on potential BSM events with the $E_\mathrm{T}^\mathrm{miss} + h(\to b\bar{b})$ final state. An overall conclusion is presented in Chap. 13.

References

1. P.A.R. Ade et al., Planck 2013 results. I. Overview of products and scientific results. Astron. Astrophys. **571**, A1 (2014)
2. D. Bauer et al., Dark matter in the coming decade: complementary paths to discovery and beyond. Phys. Dark Univ. **7–8**, 16–23 (2015)
3. G. Steigman, M.S. Turner, Cosmological constraints on the properties of weakly interacting massive particles. Nucl. Phys. **B253**, 375 (1985)
4. D. Abercrombie et al., Dark matter benchmark models for early LHC Run-2 searches: report of the ATLAS/CMS dark matter forum. Submitted to Phys. Dark Univ. (2015). arXiv:1507.00966 [hep-ex]
5. J. Abdallah et al., Simplified models for dark matter searches at the LHC. Phys. Dark Univ. **9–10**, 8–23 (2015)
6. ATLAS Collaboration, Search for new phenomena in final states with an energetic jet and large missing transverse momentum in pp collisions at $\sqrt{s} = 8$ TeV with the ATLAS detector. Eur. Phys. J. **C75**, 299 (2015)
7. CMS Collaboration, Search for dark matter, extra dimensions, and unparticles in monojet events in proton–proton collisions at $\sqrt{s} = 8$ TeV. Eur. Phys. J. **C75**, 235 (2015)
8. ATLAS Collaboration, Search for dark matter in events with heavy quarks and missing transverse momentum in pp collisions with the ATLAS detector. Eur. Phys. J. **C75**, 92 (2015)
9. CMS Collaboration, Search for monotop signatures in proton-proton collisions at $\sqrt{s} = 8$ TeV. Phys. Rev. Lett. **114**(10), 101801 (2015)
10. ATLAS Collaboration, Search for new phenomena in events with a photon and missing transverse momentum in pp collisions at $\sqrt{s} = 8$ TeV with the ATLAS detector. Phys. Rev. **D91**, 012008 (2015)
11. CMS Collaboration, Search for dark matter and large extra dimensions in pp collisions yielding a photon and missing transverse energy. Phys. Rev. Lett. **108**, 261803 (2012)
12. ATLAS Collaboration, Search for dark matter in events with a Z boson and missing transverse momentum in pp collisions at $\sqrt{s} = 8$ TeV with the ATLAS detector. Phys. Rev. **D90**, 012004 (2014)

13. ATLAS Collaboration, Search for dark matter in events with a hadronically decaying W or Z boson and missing transverse momentum in pp collisions at $\sqrt{s} = 8\,\text{TeV}$ with the ATLAS detector. Phys. Rev. Lett. **112**, 041802 (2014)
14. ATLAS Collaboration, Observation of a new particle in the search for the Standard Model Higgs boson with the ATLAS detector at the LHC. Phys. Lett. B **716**, 1–29 (2012)
15. CMS Collaboration, Observation of a new boson at a mass of 125 GeV with the CMS experiment at the LHC. Phys. Lett. B **716**, 30–61 (2012)
16. L. Carpenter et al., Mono-Higgs-boson: a new collider probe of dark matter. Phys. Rev. **D89**, 075017 (2014)
17. A. Berlin, T. Lin, L.-T. Wang, Mono-Higgs detection of dark matter at the LHC. J. High Energy Phys. **06**, 078 (2014)
18. ATLAS Collaboration, Search for dark matter in events with missing transverse momentum and a Higgs boson decaying to two photons in pp collisions at $\sqrt{s} = 8\,\text{TeV}$ with the ATLAS detector. Phys. Rev. Lett. **115**(13), 131801 (2015)

Chapter 2
The Standard Model and Beyond

The Standard Model (SM) is a theory that describes the fundamental particles and their interactions, with the exception of gravitational interactions. Experimental observations over the past several decades have proven that the SM accurately describes the physics at energy levels accessible in the laboratory. The discovery of the Higgs boson during RunI of the LHC marks the completion of the theory, but various experimental hints and theoretical calculations also point to limitations of the SM.

This chapter is organized as follows: Sect. 2.1 starts by providing an overview of the Standard Model; the theory of the Higgs boson and its discovery is described in Sect. 2.2, and Sect. 2.3 sketches out the open questions and possible answers regarding physics beyond-the-Standard-Model that are most pertinent to the topics of this thesis.

2.1 The Standard Model

The Standard Model (SM) [1–4] of particle physics has been developed through the latter half of the twentieth century, guided by both theoretical predictions and experimental discoveries. It encompasses three of the four fundamental forces of nature with the exception of gravity, namely the strong, electromagnetic, and weak forces. In the SM, all matter is made up of two types of particles: fermions that have half-integer spin and are governed by Fermi–Dirac statistics, and bosons that have integer spin and follow Bose–Einstein statistics. The fundamental building blocks of matter are fermions in the SM, while the mediators of forces that dictate their interactions are bosons. Each particle in the SM, whether a fermion or a boson, has a corresponding antiparticle identical in mass but opposite in quantum charge: in most cases a particle and its antiparticle are different particles, but occasionally a

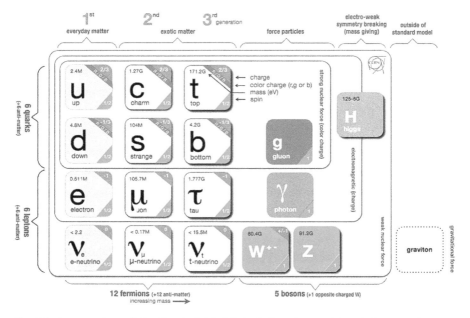

Fig. 2.1 An overview of the Standard Model, including the particles, their properties, and interactions

particle may be its own antiparticle. Figure 2.1 summarizes the different particles in the SM, their properties, and interactions.

2.1.1 Bosons

The different groups of bosons in the SM act as mediators for each of the three types of force described in the SM.

The photon is the mediator of the electromagnetic force, and couples to all fermions with a non-zero electromagnetic charge. The photon itself is massless, electromagnetic-charge-neutral, and has a spin of 1. The photon is its own antiparticle and was the first boson to be examined experimentally.

The gluon is the mediator of the strong force, and couples to all fermions with a color charge. The gluon carries color charge itself as well, which means unlike the photon, the gluon not only mediates the strong force, but also participates in it. Depending on the different combinations of the color charge, gluons come in eight varieties, given that the ninth possibility in the form of the singlet state $(r\bar{r} + b\bar{b} + g\bar{g})/\sqrt{3}$ does not exist, indicating that strong interactions happen on short distance scales. Gluons are massless and have a spin of 1. They were first discovered at DESY in the late 1970s.

2.1 The Standard Model

The $W^{\pm}$ and Z bosons are the mediators of the weak force, and couple to all fermions. The $W^{\pm}$ bosons carry the weak charged current, while the Z boson is the mediator of the weak neutral current. The $W^{\pm}$ bosons have electromagnetic charges of ± 1, and are each other's antiparticle. The Z boson is electromagnetic-charge-neutral and its own antiparticle. The $W^{\pm}$ and Z bosons have a spin of 1. They were discovered at the UA1 and UA2 experiments at CERN in the late 1980s.

The final boson, which was also the last missing piece of the SM until its discovery at the LHC in 2012, is the Higgs boson, or the boson of the Brout–Englert–Higgs mechanism. Its properties and discovery will be discussed in more detail in Sect. 2.2.

2.1.2 Fermions

The fermions in the SM are divided into two groups of very different properties, namely quarks and leptons, all of which have spin of 1/2. Both quarks and leptons can be divided into three generations, where the first generation corresponds to what exists in common matter, and the second and third generations, having higher masses, can be accessed at increasingly higher energies. The quarks feel the strong force, while the leptons do not.

Quarks, regardless of the generation, can be divided into two types depending on the electromagnetic charge. The *up-type* quarks have a charge of $2/3$, and the *down-type* quarks have a charge of $-1/3$. Leptons can be categorized as charged leptons and neutrinos.

2.1.3 Gauge Theory

The SM is based on a mathematical framework called quantum field theory [5], which is used to construct quantum mechanical models of particles. Within the framework, particles are represented by states of quantized fields. Interactions among particles are governed by a Lagrangian. The SM is a gauge theory, meaning the Lagrangian is invariant under a continuous group of local transformations. In particular, the SM Lagrangian is invariant under transformations of the group $SU(3)_c \times SU(2)_L \times U(1)_Y$.

Gauge fields with integer spin are included in the model to maintain this invariance, and excitations in these fields correspond to gauge bosons. There are a total of twelve gauge bosons: eight gluons corresponding to the generators of $SU(3)_c$, two oppositely charged W bosons corresponding to generators of $SU(2)_L$, and a neutral Z boson and a photon (γ), which correspond to linear combinations of generators for $SU(2)_L$ and $U(1)_Y$. The gauge bosons ensure that the SM is renormalizable, which is a form of consistency that is necessary for the model to have predictive power.

To match the gauge theory, the left-handed fermions exist as doublets under $SU(2)_L$, while the right-handed fermions are singlets. This gives rise to the three generations of quark doublets and singlets, as shown in Eq. (2.1), as well as the three generations of lepton doublets and singlets, with the caveat that there are no right-handed neutrinos or left-handed anti-neutrinos [Eq. (2.2)].

$$\begin{pmatrix} u \\ d' \end{pmatrix}_L \quad \begin{pmatrix} c \\ s' \end{pmatrix}_L \quad \begin{pmatrix} t \\ b' \end{pmatrix}_L \quad u_R \quad d_R \quad c_R \quad s_R \quad t_R \quad b_R \tag{2.1}$$

$$\begin{pmatrix} \nu_e \\ e^- \end{pmatrix}_L \quad \begin{pmatrix} \nu_\mu \\ \mu^- \end{pmatrix}_L \quad \begin{pmatrix} \nu_\tau \\ \tau^- \end{pmatrix}_L \quad \overline{e_R} \quad \overline{\mu_R} \quad \overline{\tau_R} \tag{2.2}$$

The quarks which are primed are *weak eigenstates* related to *mass eigenstates* by the Cabibbo–Kobayashi–Maskawa (CKM) matrix [Eq. (2.3)].

$$\begin{pmatrix} d' \\ s' \\ b' \end{pmatrix} = \begin{pmatrix} V_{ud} & V_{us} & V_{ub} \\ V_{cd} & V_{cs} & V_{cb} \\ V_{td} & V_{ts} & V_{tb} \end{pmatrix} \begin{pmatrix} d \\ s \\ b \end{pmatrix} = \hat{V}_{\text{CKM}} \begin{pmatrix} d \\ s \\ b \end{pmatrix}. \tag{2.3}$$

The diagonal elements in the CKM matrix [Eq. (2.3)] are nearly 1, showing the dominance of the same generation. However, the non-zero off-diagonal elements indicate the possibility of generation-changing and flavor-changing processes, mediated by the $W^\pm$ bosons unique to the weak force.

2.2 Spontaneous Symmetry Breaking and the Higgs Boson

The theory described in Sect. 2.1 presents a fairly good framework of particles and their interactions; however, it also poses a number of problems when compared with experimental observations, especially in terms of the masses of the particles discovered. First, consider the part of the SM that describes the electromagnetic and weak interactions, governed by the $SU(2)_L \times U(1)_Y$ symmetry. To preserve gauge invariance the gauge bosons need to be massless (i.e., the gauge fields must be included without mass terms). However, the $W^\pm$ and Z bosons responsible for mediating weak interactions need to have large masses to properly describe the weak force. The masses of the quarks and leptons show another problem. Weak interactions are found to violate parity, coupling differently to left- and right-handed fermion helicity states. This is accounted for in the SM by treating left- and right-handed fermions as different fields with different couplings. A fermion mass term in the Lagrangian would couple these different fields, hence breaking the gauge invariance. Thus, the fermions should be massless as implied by a gauge invariant left-handed interaction, which is inconsistent with observations.

These problems can be resolved by the mechanism called "spontaneous symmetry breaking" [6–10]. Additional quantum fields with zero-spin (scalar) that couples to the $SU(2)_L \times U(1)_Y$ electroweak gauge fields are added to the Lagrangian. The scalar fields are constructed such that the zero values do not correspond to the lowest energy state. Instead, the potential takes on the shape of a *Mexican Hat*. Starting from the origin which has a local maximum, the potential drops to the minimum before rising again, and the local minimum is identical around the ring of the local potential. As a result, while the Lagrangian preserves gauge invariance under $SU(2)_L \times U(1)_Y$, the symmetry is broken in the ground state and the scalar fields take on a non-zero value, referred to as the vacuum expectation value, or *vev*. The *vev* couples to fermion and gauge fields while preserving gauge invariance, giving rise to the masses of gauge bosons and fermions observed in nature.

Spontaneous symmetry breaking also predicts a neutral, massive scalar boson called the Higgs boson h. The couplings of the Higgs to gauge bosons in the SM are determined by the gauge couplings, and the couplings to fermions are proportional to the fermion masses. The mass of the Higgs itself depends on an arbitrary parameter associated with the symmetry breaking and thus is not fixed in the SM.

At the time of the turn-on of the LHC, the Higgs boson was the last missing piece of the SM. The monumental moment in the search for the SM Higgs boson came on July 4, 2012, when the ATLAS and CMS collaborations both announced discoveries of a new "SM Higgs–like" particle with a mass near 125 GeV [11, 12]. Extensive experimental studies of the properties of the new boson, and how it interacts with other particles ensued, all of which yielding results consistent with SM predictions (Fig. 2.2), including its couplings to fermions and bosons, production rate, decay width, spin, and CP quantum numbers. With the increase in energy and luminosity at RunII of the LHC resulting in orders of magnitude larger production rates, future measurements of the Higgs properties could reveal deviations from the SM.

2.3 Physics Beyond the Standard Model

Despite being one of the most robust and successful physics models in describing the world we know of, the SM has its limitations. Increasing experimental evidence and theoretical conjectures point to a widely accepted view that a more fundamental theory exists with the SM being its low-energy realization.

One open question is the striking asymmetry of matter versus antimatter. Basic laws of symmetry and thermal equilibrium should have resulted in the production of an equal amount of matter and antimatter at the creation of the universe. However, the visible universe today is dominated by matter while antimatter has essentially vanished. This is known as Charge–Parity–Violation (CPV). While the SM provides CPV terms that can account for part of the asymmetry, it is an open topic for investigation as whether the parameters are sufficient to present the level of matter–antimatter asymmetry in the current universe.

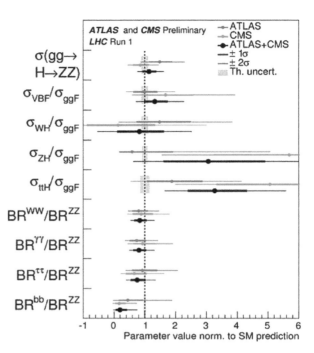

Fig. 2.2 Best-fit values of the $\sigma(gg \to H \to ZZ)$ cross-section and of ratios of cross-sections and branching ratios for the combination of ATLAS and CMS measurements of the Higgs boson. Also shown for completeness are the results for each experiment. The *error bars* indicate the 1σ (*thick lines*) and 2σ (*thin lines*) intervals. In this figure, the fit results are normalized to the SM predictions for the various parameters and the *shaded bands* indicate the theory uncertainties on these predictions

The phenomenon of neutrino oscillations, observed in both atmospheric neutrinos originating from electromagnetic cascades initiated by cosmic rays, and solar neutrinos, poses another serious challenge. This indicates that neutrinos do have mass and the flavor of the neutrino may oscillate under a rotation matrix (the PMNS matrix) similar to the CKM matrix for quarks, while the SM stipulates that neutrinos are massless.

As mentioned in Sect. 2.2, the discovery of the Higgs boson completes the SM, though the SM gives no prediction of the Higgs boson mass, which is currently measured to be $125.09 \pm 0.21(\text{stat}) \pm 0.11(\text{syst})$ GeV [13] (the "stat" denotes the statistical uncertainty and "syst" refers to the systematic uncertainty). This means the electroweak scale is $\mathcal{O}(100\,\text{GeV})$, while the Planck scale is at $\mathcal{O}(10^{19}\,\text{GeV})$. The enormous difference is called the *hierarchy problem*, indicating that either the universe is incredibly fine-tuned, or there is some form of new physics that can cancel out the divergence and stabilize the Higgs boson mass. Another fundamental question is the so-called *grand unification*, concerning whether there is a unified description addressing the three gauge interactions in the SM, i.e., electromagnetic, weak, and strong interactions. How gravity might be able to be merged into a greater symmetry adds another layer of mystery to our understanding of the universe.

Finally, the SM provides no viable candidate to account for the abundance of DM in the universe, let alone the case of dark energy. The neutrinos, despite being electromagnetically neutral and weakly interacting particles with non-zero masses, have been essentially ruled out as dominant DM candidates because they are simply not abundant enough. The existence of DM is hence a strong motivation to search for physics beyond the SM (BSM). Various BSM theories predict new particles that may be DM candidates, which can be tested through experiments that in turn constrain such theories.

References

1. S.L. Glashow, Partial symmetries of weak interactions. Nucl. Phys. **22**, 579–588 (1961)
2. S. Weinberg, A model of leptons. Phys. Rev. Lett. **19**, 1264–1266 (1967)
3. A. Salam, Weak and electromagnetic interactions. Conf. Proc. **C680519**, 367–377 (1968)
4. G. 't Hooft, M.J.G. Veltman, Regularization and renormalization of gauge fields. Nucl. Phys. B **44**, 189–213 (1972)
5. S. Weinberg, *The Quantum Theory of Fields*. Foundations, vol. 1 (Cambridge University Press, Cambridge, 1995)
6. P.W. Higgs, Broken symmetries, massless particles and gauge fields. Phys. Lett. **12**, 132–133 (1964)
7. P.W. Higgs, Broken symmetries and the masses of gauge bosons. Phys. Rev. Lett. **13**, 508–509 (1964)
8. F. Englert, R. Brout, Broken symmetry and the mass of gauge vector mesons. Phys. Rev. Lett. **13**, 321–323 (1964)
9. G.S. Guralnik, C.R. Hagen, T.W.B. Kibble, Global conservation laws and massless particles. Phys. Rev. Lett. **13**, 585–587 (1964)
10. P.W. Higgs, Spontaneous symmetry breakdown without massless bosons. Phys. Rev. **145**, 1156–1163 (1966)
11. ATLAS Collaboration, Observation of a new particle in the search for the Standard Model Higgs boson with the ATLAS detector at the LHC. Phys. Lett. B **716**, 1–29 (2012)
12. CMS Collaboration, Observation of a new boson at a mass of 125 GeV with the CMS experiment at the LHC. Phys. Lett. B **716**, 30–61 (2012)
13. ATLAS and CMS Collaborations, Combined measurement of the Higgs boson mass in pp collisions at $\sqrt{s} = 7$ and 8 TeV with the ATLAS and CMS experiments. Phys. Rev. Lett. **114**, 191803 (2015)

Chapter 3
An Overview of Dark Matter

Dark matter (DM), the existence of which was observed through the gravitational effects in large astrophysical systems, is one of the greatest mysteries in physics today. In the Standard Model discussed in Chap. 2, there is no viable dark matter candidate. This chapter is devoted to give an overview of what we know so far of dark matter, and the theoretical and experimental efforts underway to find out more about this elusive matter that makes up more than a quarter of the mass–energy of our universe.

This chapter is organized as follows: Sect. 3.1 gives a brief overview of the standard cosmological model; Sect. 3.2 discusses the calculation of the abundance of DM in our universe based on its cosmological history and thermal equilibrium rules; the compelling evidence for the existence of DM at different astrophysical scales is presented in Sect. 3.3; Sect. 3.4 lists a few of the leading hypothesis of DM candidates and their physical laws; finally, the experimental methods, categorized as direct detection, indirect detection, and collider production, are described in Sect. 3.5.

3.1 Standard Cosmology

The so-called *Big Bang* scenario, which describes the Universe as a system evolving from a highly compressed state existing around 10^{10} years ago, is a generally accepted theory underlining the *Standard* cosmological model. This model, albeit partial, allows us to explain in a satisfactory way many observed properties of the Universe, including the thermal history, relic background radiation, abundance of elements, and large scale structures.

The standard cosmological model starts from the Einstein equation [Eq. (3.1)], which describes the relationship between the geometry of the universe and its mass and energy content.

$$R_{\mu\nu} - \frac{1}{2}g_{\mu\nu}R = -\frac{8\pi G_N}{c^4}T_{\mu\nu} + \Lambda g_{\mu\nu}, \qquad (3.1)$$

The left side of the equation describes the geometry of the universe, and the right side defines its energy and momentum. where $R_{\mu\nu}$ and R are, respectively, the Ricci tensor and scalar (obtained by contraction of the Riemann curvature tensor). $g_{\mu\nu}$ is the metric tensor, G_N is Newton's constant, $T_{\mu\nu}$ is the energy-momentum tensor, and Λ is the so-called cosmological constant.

The cosmological constant term was initially introduced by Einstein to obtain a stationary solution for the Universe and subsequently abandoned when the expansion of the Universe was discovered. Its importance is only realized later, as the presence of the constant Λ indicates that in the absence of any energy or momentum associated with the matter content of the universe, i.e., $T_{\mu\nu} = 0$, there is a a "vacuum energy" associated with space-time itself, and is a source of gravitational field even in the absence of matter.

With the Einstein equation defined, one needs to specify the metrics, i.e., the symmetries of the problem in order to solve the equation. Here we assume the properties of statistical *homogeneity* and *isotropy* of the Universe, which not only greatly simplifies the mathematical analysis, but are also confirmed by many observations, for example, the isotropy of the cosmic microwave background (CMB), and a homogeneous distribution at scales in excess of $\sim$100 Mpc from galaxy surveys. With this metric, the line element can be expressed as

$$ds^2 = -c^2dt^2 + a(t)^2 \left(\frac{dr^2}{1 - kr^2} + r^2 d\Omega^2 \right), \qquad (3.2)$$

where $a(t)$ is the so-called *scale factor* and the constant k, describing the spatial curvature, can take the values $k = -1, 0, +1$, where, for example, $k = 0$ would imply a flat universe.

Solving the Einstein equations with this metric leads to the Friedmann equation

$$\left(\frac{\dot{a}}{a}\right)^2 + \frac{k}{a^2} = \frac{8\pi G_N}{3}\rho_{\text{tot}}, \qquad (3.3)$$

where ρ_{tot} is the total average energy density of the universe, and $\frac{\dot{a}(t)}{a(t)}$ is the Hubble parameter $H_0 = 73 \pm 3 \, \text{km} \, \text{s}^{-1} \, \text{Mpc}^{-1}$.

3.1 Standard Cosmology

From Eq. (3.3), the critical density ρ_c is the energy density with a flat universe, i.e., $k = 0$:

$$\rho_c \equiv \frac{3H^2}{8\pi G_N} \ . \tag{3.4}$$

By convention, the abundance of a substance i in the Universe, Ω_i, can be expressed as ratio of its density ρ_i over ρ_c:

$$\Omega_i \equiv \frac{\rho_i}{\rho_c} \ . \tag{3.5}$$

And the total density of the universe is the sum of Ω_i:

$$\Omega = \sum_i \Omega_i \equiv \sum_i \frac{\rho_i}{\rho_c}, \tag{3.6}$$

The Friedmann equation [Eq. (3.3)] can thus be written as

$$\Omega - 1 = \frac{k}{H^2 a^2} \ . \tag{3.7}$$

The sign of k is therefore determined by the value of Ω with respect to 1. The shape of the universe based on the values of the average density with respect to the critical density is given in Table 3.1.

After defining the Einstein equation [Eq. (3.1)] and the metrics, the third fundamental piece in the standard cosmological model is the equation of state describing the evolution over time. The expansion rate, which depends on the state of individual components, can be expressed as:

$$\frac{H^2(z)}{H_0^2} = \left[\Omega_X (1+z)^{3(1+\alpha_X)} + \Omega_K (1+z)^2 + \Omega_M (1+z)^3 + \Omega_R (1+z)^4 \right]$$

where M and R are labels for matter and radiation, $\Omega_K = \frac{-k}{a_0^2 H_0^2}$ and X refers to a generic substance with equation of state $p_X = \alpha_X \rho_X$ (in particular, for the cosmological constant, $\alpha_\Lambda = -1$), and z is the redshift.

Table 3.1 Classification of cosmological models based on the value of the average density, ρ, in terms of the critical density, ρ_c

$\rho < \rho_c$	$\Omega < 1$	$k = -1$	Open
$\rho = \rho_c$	$\Omega = 1$	$k = 0$	Flat
$\rho > \rho_c$	$\Omega > 1$	$k = 1$	Closed

3.2 Relic Density

Based on the standard model of cosmology, one could trace the history of our universe by extrapolating known physics back to the Planck epoch, when the universe was $t = 10^{-43}$ s old or less with an energy scale at the Planck mass of $M_{\text{Pl}} = 10^{19}$ GeV, and gravitational interactions were either dominant or part of an overall theory. At the end of the Planck epoch, the gravitational force has lost its significance, and the strong, weak, and electromagnetic forces were likely unified into an overall theory. We omit the complete run-down of the major epochs since the Planck epoch up to the current day energy scale of $T = 2.7$ K $\sim 10^{-4}$ eV here, but only point out two periods important to the existence of DM in the universe today and its constraints.

- Variable time: this is the period when weakly interacting dark matter candidates with energy scale of $\sim 10^1 - 10^3$ GeV *freeze-out*, which can also span variable scales depending on the nature of the DM.
- $t = 0.1 \rightarrow 10^3$ s: this is the era of Big Bang Nucleosynthesis (BBN) with energy scale of ~ 100 KeV up to 10 MeV. During this period, protons and neutrons form light elements including D, ^{3}He, ^{4}He, and Li.

The standard cosmological model, also known as the ΛCDM model, is named by suggesting a non-zero cosmological constant Λ, and dark matter (DM) as a Cold (C), i.e., non-relativistic particle when the *freeze-out* or decoupling occurred. It is possible that DM is *hot*, i.e., can decouple at relativistic energies, or are not in the form of particles, but these hypothesis are strongly disfavored by empirical evidence and theoretical projections. Assuming DM is cold, one can calculate the abundance of DM in the universe today, the so-called *relic density*, based on thermal equilibrium of a system using the Boltzmann equation. As the universe expands and its temperature decreases, the ambient energy becomes insufficient to produce the particle. The existing particles continue to interact and annihilate, until the interaction rate drops below the expansion rate of the universe and the equilibrium could not long be maintained, at which point the particle is said to be *decoupled*.

The relic density can be expressed in terms of the critical density [see Eq. (3.5)]

$$\Omega_\chi h^2 \approx \frac{1.07 \times 10^9 \text{ GeV}^{-1}}{M_{\text{Pl}}} \frac{x_F}{\sqrt{g_*}} \frac{1}{(a + 3bT_F/m_\chi)}, \qquad (3.8)$$

where a and b have units of GeV^{-2}, T_F is the *freeze-out* temperature denoting the energy at which *freeze-out* occurs, g_* is evaluated at the freeze-out temperature, and $h = H_0/100$ km s^{-1} Mpc^{-1} where H_0 is the Hubble parameter. A frequently used estimation accurate within an order of magnitude is

$$x_F = \ln\left[c(c+2)\sqrt{\frac{45}{8}}\frac{g}{2\pi^3}\frac{m\, M_{\text{Pl}}(a + 6b/x_F)}{g_*^{1/2} x_F^{1/2}}\right], \qquad (3.9)$$

where c is a constant of order one determined by matching the late-time and early time solutions.

It is worth noting that the aforementioned equations are obtained with a series of simplifying assumptions. Scenarios like the presence of a scalar field in the early universe, or resonance enhancements (*coannihilations*) can drastically change the value of the relic density. In particular, the case of coannihilations is well-established in models with supersymmetric particles.

3.3 Evidence of Dark Matter

While the exact nature of DM remains a mystery, many independent observations of different astrophysical scales have made strong cases for the existence of DM in our universe.

On the galactic scale, the most convincing and direct evidence of DM comes from observing the rotation curves of galaxies. The graph of circular velocities of stars and gas inside a galaxy as a function of their distance from the galactic center, i.e., the *rotation curve*, often exhibits a characteristic flat behavior at large distances far beyond the edge of visible disks. This strongly suggests the existence of a "dark" halo with $M(r) \propto r$ and $\rho \propto 1/r^2$, though the total amount of dark matter present is difficult to quantify not knowing the distances to which the halos extend. Other evidence for DM at the galactic scale comes from mass modeling of the detailed rotation curves, and strong gravitational lensing in some elliptical galaxies.

Moving up the observational scale, we look at clusters of galaxies. For the baryonic mass of a typical cluster, the temperature should follow the equation

$$kT \approx (1.3 - 1.8)\,\text{keV} \left(\frac{M_r}{10^{14} M_\odot}\right)\left(\frac{1\,\text{Mpc}}{r}\right) \quad (3.10)$$

where M_r is the baryonic mass enclosed within the radius r. The disparity between the temperature obtained using Eq. (3.10) and the corresponding observed temperature, $T \approx 10\,\text{keV}$, suggests the existence of a substantial amount of dark matter in galactic clusters, which may be compared against estimates from gravitational lensing data.

Scanning the entire universe, crucial information on both the existence and the amount of DM can be obtained from analysis of the CMB, which is the background radiation from the propagation of photons in the early universe, once they were decoupled from matter. CMB is experimentally measured to be isotropic at the 10^{-5} level and appears to follow with extraordinary precision the spectrum of a black body corresponding to a temperature $T = 2.726\,\text{K}$. By extracting information from CMB anisotropy maps, the abundance of baryons and matter in the Universe is found to be (Spergel et al. [1])

$$\Omega_b h^2 = 0.0224 \pm 0.0009 \quad \text{and} \quad \Omega_M h^2 = 0.135^{+0.008}_{-0.009}. \quad (3.11)$$

The value of $\Omega_b h^2$ thus obtained is consistent with predictions from BBN (Olive [2])

$$0.018 < \Omega_b h^2 < 0.023. \tag{3.12}$$

Finally, we bring the focus back to our galaxy, the Milky Way, where there is an abundance of observational data across wide ranges. Observations of the velocity dispersion of high proper motion stars suggest the existence of a super massive black hole (SMBH) lying at the center of our galaxy, with a mass, $M_{\text{SMBH}} \approx 2.6 \times 10^6 M_\odot$ [3], the process of adiabatic accretion of DM on which would produce "spikes" in the DM density profile. These "spikes" would in turn produce dramatic increases in the annihilation radiation from the galactic center, which may be observed through indirect detection experiments introduced in Sect. 3.5. The local density of DM can also be determined by analyzing the rotation curves of the Milky Way.

3.4 Theories of Dark Matter Candidates

Despite the compelling evidence for DM as presented in the previous section, we have yet to observe any coupling to DM with a force other than gravity, which is too weak for localized experimental searches. Thus, a plethora of theoretical models have been proposed to describe the possible nature of DM and its interaction with SM observables, the most compelling of which being that DM is a type of weakly interacting massive particle (WIMP). The so-called WIMP miracle calculates that a particle with a mass at the weak scale ($\mathcal{O}(\text{GeV})$ to $\mathcal{O}(\text{TeV})$) which has a weak coupling to the SM can match the observed relic density from CMB analysis.

Of the WIMP candidates of DM particles, two classes are particularly interesting, as both also belong to popular extensions of the Standard Model, namely supersymmetry (SUSY) and extra-dimensions.

Supersymmetry As a popular proposed solution to the hierarchy problem, among other theoretical motivations, SUSY introduces a fermionic "Superpartner" to each boson, and vice versa. Different SUSY models of varying degrees of complexity and completeness exist, starting from the minimal supersymmetric extension of the Standard Model (MSSM), where fermionic superpartners are associated with all gauge fields, scalar partners are associated with fermions, and one additional Higgs field is introduced yielding a total of two Higgs doublets corresponding to five Higgs bosons.

One feature of the MSSM is the conservation of R-parity. R-parity is a multiplicative quantum number defined as

$$R \equiv (-1)^{3B+L+2S}. \tag{3.13}$$

All of the Standard Model particles have R-parity $R = 1$ and all "sparticles" (i.e., superpartners) have $R = -1$. Due to R-parity conservation, sparticles can only decay into an odd number of sparticles plus Standard Model particles. The lightest sparticle (lightest supersymmetric particle, or LSP) hence is stable and can only be destroyed via pair annihilation, making it an excellent DM candidate. Among potential LSP particles, the neutralino, which is a Majorana fermionic mass eigenstate from mixing of the spartners of the photon, Z, and the neutral Higgs bosons, is not excluded by direct detection experiments, and can be sensitive to searches via its self-annihilation or interactions with SM particles. There have been a number of additional sparticles proposed by different SUSY models which can be DM candidates.

Extra-Dimensions In the search for a fundamental unification theory encompassing all known interactions, theories with extra spatial dimensions are developed. In models with unified extra-dimensions, in which all particles and fields of the SM can propagate into extra-dimensions, the lightest of all states corresponding to the first excitations of the SM particles, i.e., the lightest Kaluza–Klein particle, is a viable DM candidate if it is electrically neutral (e.g., Cheng et al. [4]).

While WIMPs remain the most appealing DM candidate, other types of DM candidates have been proposed as well. One of them is *Axions* (e.g., Duffy and van Bibber [5]), which are expected to be very light on the order of 0.01 eV or below, and extremely weakly interacting. More recently, there are proposals of DM being strongly interacting massive particles (SIMPs, e.g., Hochberg et al. [6]), or residing inside a large family of "dark sector" particles.

It is important to keep in mind that there is no proof that DM is made of a single particle species. Quite the contrary, we already know SM neutrinos contribute to DM, but are simply not abundant enough. While the upper bound on relic density from CMB analysis is a strict limit, the lower bound can be relaxed assuming the target particle of the search, despite being a viable DM candidate, is only a sub-dominant component of all DM in the universe. This assumption is particularly relevant in experimental searches for DM, as will be discussed in the next section.

3.5 Experimental Methods

As discussed in Sect. 3.4, there are many types of theoretical models for DM. Depending on the underlying theoretical assumptions, DM particles (assuming they have particle form) can interact with SM particles in drastically different ways, making searches for DM more model-dependent. For the focus of this thesis, as is with the leading hypothesis, we assume DM particles are WIMPs. Other classes of DM could potentially be detected with similar methods, but they are not discussed here. Under the WIMP assumption, there are two generic types of DM searches. The first is direct searches for particles in ultra-violet (UV) complete models, like SUSY

Fig. 3.1 The three categories of experimental methods to search for WIMP-type dark matter. Direct detection looks for DM particle scattering off a SM particle; indirect detection measures possible annihilation of DM particles into SM particles; and a collider experiment searches for DM particle pairs produced from SM particle interactions. The three approaches are complementary

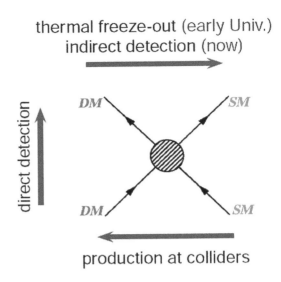

searches at collider experiments, where one or more particles in the model may be a viable DM candidate, but the discovery of the particle is not be the main motivation of the search. This approach is entirely model-dependent, and is also not discussed further. The second classification pertains to more generic DM searches, where there is a "mediator" connecting the SM to the dark sector, as seen in Fig. 3.1. This leads to three related Feynman diagrams that can be obtained by rotating Fig. 3.1 as appropriate, implying all three approaches have the ability to probe the same type of interaction between DM and SM particles. On the other hand, this figure is a generalization, and in actual experiments, the three approaches may be searching for different types of DM particles.

3.5.1 Direct Detection

Direct detection is one of the most promising methods to search for DM particles, with the ability to probe a wide range of DM mass. The basic idea is that if our universe is filled with DM particles, then it is highly likely some of them may travel through the atmosphere and reach earth, where they would interact with SM particles, and one could build experiments to detect signals of such an interaction. Direct detection experiments make the assumption that WIMPs from the galaxies are non-relativistic with velocity similar to the velocity of stars within the Milky Way $\sim \mathcal{O}(100\,\text{km/s})$, hence the typical DM-SM scattering interaction is a nuclear recoil event. For WIMP masses on the GeV to TeV scale, the energy transfer in such an event, assuming the DM scattering is elastic, is at the level of $\mathcal{O}(1\,\text{KeV})$ to $\mathcal{O}(100\,\text{KeV})$, which may be detected through ionization, scintillation,

3.5 Experimental Methods   21

or phonon methods. To reduce background, direct detection experiments are usually deep underground, and employ special detector shielding to reject background from radioactive decays.

Direct detection experiments can be further categorized as SI (spin-independent) types, where the interaction cross-section only depends on the number of nucleons and motivates the usage of massive target particles like germanium or xenon, and SD (spin-dependent) experiments, where the coupling between the WIMP and the target SM particle depends on the nuclear spin, motivating the usage of highly polarized target material such as fluorine. Some of the leading SI experiments include CDMS-II, CRESST-II, CoGeNT, LUX, and Xenon100. For SD experiments, there are COUPP, IceCube, PICASSO, and SIMPLE.

3.5.2 Indirect Detection

Compared with direct detection experiments in the previous section, indirect detection experiments shift its gaze from deep underground to the distant skies. As indirect detection experiments search for the resulting decay products of DM annihilation, the process relies on the annihilation rate proportional to DM density, and higher DM density can be expected in regions of strong gravitation, such as centers of galaxies or within stars. The decay products may take different forms, but the most common searches are for excesses in antiparticle flux or mono-energetic lines in the photon spectrum. Leading indirect detection experiments include space-based experiments such as AMS and Fermi-LAT, both of which have recently presented results of antiparticle flux excesses which could be compatible with DM annihilation, but may also be a result of other nearby astrophysical sources like pulsars. Further analysis of data and interpretation of results are underway. There is also another class of indirect detection experiments using large telescope arrays, such as HESS, which can be used to constrain DM annihilation rate through measurements of the photon spectrum, and is shown to be typically more sensitive than space-based experiments for high mass DM particles.

3.5.3 Collider Production

Finally, DM particles can be directly produced at collider experiments, where the noninteracting DM particles escape the detector, leaving the distinct signature of large missing transverse momentum, for energy-momentum conservation, recoiling against one or more visible particles. By searching for an excess of such events, one could place constraints on DM production in association with the visible particle(s). The results can be translated to limits on the WIMP-nucleon scattering cross-section and the WIMP-WIMP annihilation cross-section, making the collider results comparable to direct detection and indirect detection experiments, respectively.

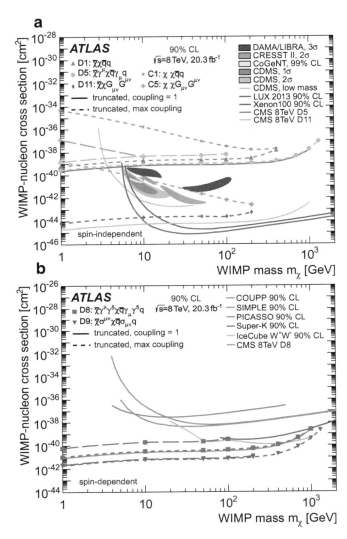

Fig. 3.2 Inferred 90 % CL limits on (**a**) the spin-independent and (**b**) spin-dependent WIMP–nucleon scattering cross-section as a function of DM mass for different operators from the ATLAS "mono-jet" search for DM (ATLAS Collaboration [7]). Results from direct detection experiments for the spin-independent and spin-dependent cross-section, and the CMS (untruncated) results are shown for comparison. Results from gamma-ray telescopes are also shown, along with the thermal relic density annihilation rate

For low mass DM particles, direct detection is challenging as the nuclear recoil energy may fall to levels below the limits of current detection technologies, typically on the scale of a few KeV. On the other hand, collider searches are usually more sensitive to low DM masses because the larger production cross-section and stronger kinematic spectrum typical for very light DM particles in comparison to heavier ones in the same model. Moreover, SD direct detection experiments are in general less sensitive than SI experiments due to the coupling to the nucleon spin rather than nucleon mass. This suppression is important at the nucleon level, but the equivalent notion of chirality at colliders has a much smaller impact. As a result, collider limits are generally stronger than both SI and SD direction detection results for low mass DM, and stronger than SD direct detection limits over the full parameter space considered. Comparisons of the results from ATLAS "mono-jet" search for DM (ATLAS collaboration [7]), and SI and SD direct detection limits, are shown in Fig. 3.2. Details on collider searches for DM, including benchmark models and search channels, will be presented in Chap. 5.

The three approaches of DM searches, direct detection, indirect detection, and collider production, are complementary to one another. It is important to combine all three methods in order to comprehensively probe the nature of DM and its interaction with the visible sector.

References

1. D.N. Spergel et al., First year Wilkinson microwave anisotropy probe (WMAP) observations: determination of cosmological parameters. Astrophys. J. Suppl. **148**, 175–194 (2003)
2. K.A. Olive, Tasi lectures on dark matter, pp. 797–851, 2003. Theoretical Advanced Study Institute in Elementary Particle Physics at the University of Colorado at Boulder, 2–28 June 2002
3. A.M. Ghez, B.L. Klein, M. Morris, E.E. Becklin, High proper motion stars in the vicinity of Sgr A*: evidence for a supermassive black hole at the center of our galaxy. Astrophys. J. **509**, 678–686 (1998)
4. H.-C. Cheng, J.L. Feng, K.T. Matchev, Kaluza-Klein dark matter. Phys. Rev. Lett. **89**, 211301 (2002)
5. L.D. Duffy, K. van Bibber, Axions as dark matter particles. New J. Phys. **11**, 105008 (2009)
6. Y. Hochberg, E. Kuflik, T. Volansky, J.G. Wacker, Mechanism for thermal relic dark matter of strongly interacting massive particles. Phys. Rev. Lett. **113**, 171301 (2014)
7. ATLAS Collaboration, Search for new phenomena in final states with an energetic jet and large missing transverse momentum in pp collisions at $\sqrt{s} = 8$ TeV with the ATLAS detector. Eur. Phys. J. **C75**, 299 (2015)

Chapter 4
The ATLAS Experiment at the Large Hadron Collider

The Large Hadron Collider (LHC) is host to a number of high energy physics experiments, one of which being the ATLAS experiment using the namesake detector, on which the work of this thesis is based. This chapter provides a brief introduction to the LHC and the ATLAS detector. More information on the LHC can be found in [1–3]. Information on the design, construction, and operation of the ATLAS detector can be found in [4–8].

This chapter is organized as follows: Sect. 4.1 gives an overview of the LHC and the data sets from its collisions used for this thesis; Sect. 4.2 provides an introduction to the ATLAS detector, including its subdetector systems; finally, the trigger and data acquisition system used to read out and process collision data recorded by the ATLAS detector for physics usage is described in Sect. 4.3, with a special note given to the Fast Tracker (FTK) track trigger upgrade that the author worked on in Sect. 4.3.2.

4.1 The Large Hadron Collider

The LHC is a super-conducting particle accelerator and collider located at the European Organization for Nuclear Research (CERN). Constructed between 1998 and 2008, the LHC is currently the largest particle collider in the world. The LHC accelerates particles around a circular 27 km tunnel buried roughly 100 m below ground across the Swiss–French border. An illustration of the machine and the main experiments on it is shown in Fig. 4.1. There are four experiments located at collision points around the LHC ring. This thesis is based on data generated by the ATLAS experiment [4], which is designed to study proton–proton (pp) collisions. The CMS experiment [9], located on the opposite side of the ring, is also designed

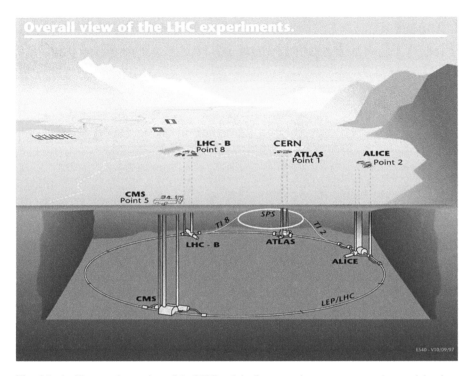

Fig. 4.1 An illustrated overview of the LHC and the four experiments constructed around the ring: ATLAS, CMS, ALICE, and LHCb

to study pp collisions. The other two experiments are ALICE [10], designed to study heavy ion collisions, and LHCb [11], designed to study physics related to b-hadrons.

Before protons are injected into the LHC, they pass through a number of smaller particle accelerators that successively increase their energy. At the beginning of the injection chain, bunches from hydrogen atoms are accelerated to an energy of 50 MeV in a linear accelerator called LINAC2. They then enter a synchrotron called the PS Booster, which accelerates them to 1.4 GeV and injects them into the Proton Synchrotron (PS). The PS accelerates the protons to 26 GeV and then passes them into the Super Proton Synchrotron (SPS). Once the protons are accelerated to 450 GeV, they are injected into the main LHC ring. This injection chain produces bunches of protons that travel in opposite directions in separate beam pipes. Superconducting RF cavities located on a section of the ring (IP4) are designed to accelerate the bunches to an energy 7 TeV. The bunches are bent around the ring by super-conducting dipole magnets that generate magnetic fields up to 8.3 T. The beam pipes are nominally designed to contain a total of 2808 bunches, each with roughly 10^{11} protons. However, since the machine has only been in full operation for 3 years it has yet to fully reach the design goals.

Designed to produce collisions with a center of mass energy of 14 TeV, the LHC was turned on for the first time in September 2008. An accident occurred shortly after it was turned on, when a quench in one of the super-conducting dipole magnets leads to high current, causing a faulty solder connection between magnets to open. An electrical arc ensued, which led to a sudden evaporation of liquid helium, and an explosion that damaged over 50 magnets. Repairs were finished in 2009. When the first LHC run (Run I) took place between 2010 and 2012, the decision was made to operate the machinery at a reduced energy until further upgrades could be made. In 2010 and 2011 the LHC ran with each beam accelerated to 3.5 TeV, producing collisions up to center of mass energy $\sqrt{s} = 7$ TeV. In 2012 the energy of the beams was increased to 4 TeV, leading to collisions up to center of mass energy $\sqrt{s} = 8$ TeV. The machine was shut down in 2013 for maintenance and upgrades. The second run (Run II) began in 2015 at center of mass energy of $\sqrt{s} = 13$ TeV.

A number of features in the proton beams used in the collisions are important to its physics potential. As already mentioned, the energy of the beam is important because higher energy collisions can be used to study more massive particles and shorter distances. Another important characteristic of the beam is luminosity. The *instantaneous* luminosity is proportional to the rate at which collisions are produced by the accelerator, so a beam with a high instantaneous luminosity is capable of producing data at a high rate. Figure 4.2 shows the peak instantaneous luminosity delivered by the LHC in 2012 [12]. The highest instantaneous luminosity was achieved in 2012 at nearly 8×10^{33} cm^{-2} s^{-1}, as compared to the design luminosity of 10^{34} cm^{-2} s^{-1}. If the instantaneous luminosity is integrated over a time window, the result is proportional to the total number of collisions. This is called the *integrated* luminosity and it is used to characterize the total size of a dataset. The number of collisions producing a specific final state, N_i, is the product of the cross-section for that process, σ_i, and the integrated luminosity, $\int L dt$. Figure 4.3 shows the integrated luminosity accumulated over time in 2012 [12]. The LHC runs in 2010 and 2011 with $\sqrt{s} = 7$ TeV generated a total integrated luminosity of roughly 5 fb^{-1}. The full run during 2012 with $\sqrt{s} = 8$ TeV generated roughly 25 fb^{-1}. The 2012 dataset was used in this thesis to search for DM particles produced in association with a Higgs boson decaying to two bottom quarks. The *pile-up* effect, referring to multiple *pp* interactions when two proton bunches cross each other, is important to take into account when generating simulated data. Figure 4.4 shows a luminosity-weighted distribution of the mean number of interactions per bunch crossing (μ) during the 2012 operation [13]. During the 2012 run the mean number of interactions per bunch crossing was as large as 40. More information on the calculation of μ can be found in [14].

4.2 The ATLAS Detector

ATLAS (A Toroidal LHC ApparatuS) is a large general-purpose particle detector at the LHC. It is designed to be sensitive to many different processes in particle physics in both *pp* collisions and heavy ion collisions, the former of which being the focus of this thesis.

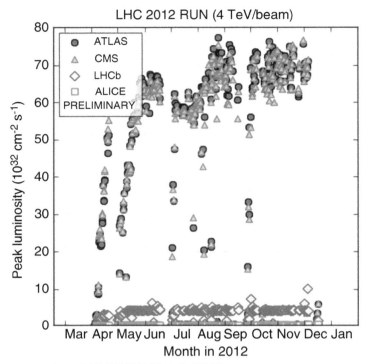

Fig. 4.2 Peak instantaneous luminosity for pp collisions delivered by the LHC in the year 2012. The periodic dips in luminosity are associated with technical stops of the machine and the introduction of new beam configurations

The ATLAS detector has a cylindrical shape and is approximately 25 m high and 44 m long, with a weight of roughly 7000 tons. The beam pipe passes through the middle of the detector along the cylindrical axis of symmetry. Collisions occur at the center and produce particles that emerge outward in all directions.

A right-handed coordinate system is used to characterize particles measured by the ATLAS detector. The z-axis (or longitudinal axis) runs along the beam line, the x-axis points towards the center of the LHC ring, and the y-axis points upward. The origin is defined as the interaction point. The $x-y$ plane perpendicular to the beam line is called the transverse plane. The momentum of particles produced in a collision is conserved in the transverse plane, so the transverse component of momentum, p_T, is a commonly used kinematic variable. Longitudinal momentum of the colliding proton constituents is not known because of the undetected produced particles that travel down the evacuated beam pipe, so the conservation of longitudinal momentum is less useful.

Since the ATLAS detector has a cylindrical shape, the Cartesian axes defined above are often converted to an $r - \phi$ coordinate system. In this case, ϕ is the azimuthal angle defined with respect to the x-axis, and r is the transverse distance

4.2 The ATLAS Detector

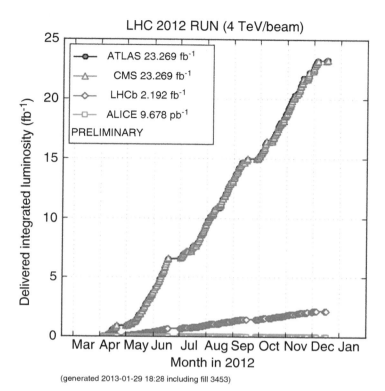

Fig. 4.3 Integrated luminosity for *pp* collisions accumulated at the LHC in the year 2012

Fig. 4.4 The luminosity-weighted distribution of the mean number of interactions per proton bunch (μ) crossing during the 2012 LHC run

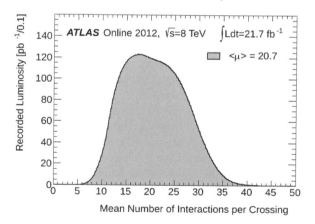

from the beam line. The polar angle θ with respect to the z-axis is often reported in terms of an approximate Lorentz invariant parameter called pseudorapidity. The pseudorapidity is given by $\eta = -\ln[\tan(\theta/2)]$. In the $\eta - \phi$ space, the distance $\Delta R = \sqrt{\Delta\eta^2 + \Delta\phi^2}$ is an approximate Lorentz invariant quantity sometimes used to characterize the separation between particles traversing through the detector.

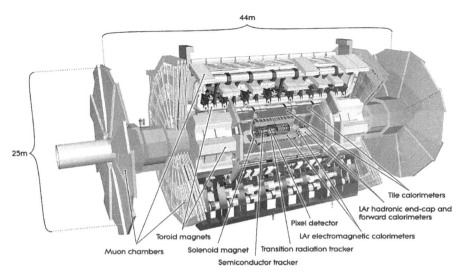

Fig. 4.5 Diagram of the ATLAS detector [4]

ATLAS records the paths and energies of the particles emerging from the collisions. Depending on how much they interact or how far they travel inside the detector, the particles produced from the collision point and their decay products may pass through several different layers of subdetectors. Each subdetector system is designed to be sensitive to different types of particles or their kinematic properties. Figure 4.5 shows an illustration of the ATLAS detector and its different subdetector systems. The innermost layer is called the Inner Detector (ID). This sub-system is designed to reconstruct the trajectories of charged particles. The ID is surrounded by the Electromagnetic and Hadronic Calorimeters, which are designed to measure the energies of electrons, positrons, photons, and hadrons. Surrounding the calorimeters is the Muon Spectrometer (MS) inside a 0.5 T toroidal magnetic field, designed to reconstruct muon trajectories.

Many subdetectors involve two types of components, designed to maximize the precision of measurements over a wide range of angles. In the central η region, also called the barrel region, where particles are approximately perpendicular to direction of the beam, long cylindrical detectors around the beam pipe are used. In the forward η region, also called the end cap region, the particle trajectory becomes more aligned with the beam, hence wheel-shaped end cap detectors are used with the axle of the wheel being the beam pipe. The two scenarios are associated with very different levels of particle flux, which often requires different detector materials. This split between barrel and end cap subdetectors applies to all three types of ATLAS detector classifications, while the exact η separation between the two differs. This classification excludes luminosity detectors and beam conditions monitors, where the former is in the very forward region and the latter exists in multiple regions.

4.2 The ATLAS Detector

A detailed description of the ATLAS detector and sub-systems is provided in [4]. The remainder of this section gives a brief overview of each subdetector system.

4.2.1 Inner Detector

The Inner Detector (ID) is the closest to the interaction point, and is used to reconstruct trajectories of charged particles with energies of at least 500 MeV. The detector consists of modular sensors built around the beam pipe with cylindrical symmetry. There are barrel layers centered at the interaction point, in addition to end cap wheels at each end of the barrel that extend the η range. The positions of charged particles are measured with high precision as they traverse the ID layers. All of the ID sensors sit inside a 2 T magnetic field aligned with the beam line, which is generated by a super-conducting solenoid. Charged particles follow helical trajectories as they pass through this field, and the curvature of the trajectories can be used for charge identification and momentum determination. The ID is further composed of three sub-systems, as illustrated in Fig. 4.6.

The innermost sub-system is called the Pixel detector [15, 16]. This component has three layers in the barrel and three layers in each end cap, providing uniform coverage in ϕ for $|\eta| < 2.5$. Each of the layers consists of many silicon sensors or pixels, each 50 by 400 μm. When charged particles pass through the silicon sensors they generate electron-hole pairs that drift in an applied electric field to a readout device. The system contains a total of 80 million readout channels and provides a position resolution of 10 μm in the $r - \phi$ plane, and 115 μm in the z-direction. This level of precision is necessary for resolving dense tracking environments, when many charged particles are produced in collisions.

The Pixel detector is surrounded by the semiconductor tracker (SCT) [17, 18], which consists of four double layers in the barrel and nine layers in each end cap, providing coverage for $|\eta| < 2.5$. The SCT is composed of strips of solid silicon that operate similarly to the silicon sensors in the Pixel detector. Each strip has a length of 126 mm and a pitch of 80 μm (in the barrel region). The SCT layers are each composed of two sets of silicon strips with axes tilted by 40 mrad with respect to one another. The entire SCT has roughly 60 million readout channels. The pairs of strips locate charged particle positions with an accuracy of 17 μm in the $r - \phi$ plane, and 580 μm along the z-axis.

The outermost sub-system in the ID is called the transition radiation tracker (TRT) [19]. The TRT contains roughly 300,000 straw drift tubes that occupy 70 layers in the barrel and 140 layers in each end cap, providing coverage for $|\eta| < 2.0$. Each straw drift tube contains a gas that is ionized by incident charged particles. The emitted electrons drift in an applied electric field to a wire that runs down the center of the straw to a readout device. A single charged particle will typically register hits in roughly 35 of the straw drift tubes, allowing the particle position to be measured with an accuracy of roughly 130 μm in the $r - \phi$ plane. In addition, the TRT is designed to help with particle identification. As charged particles pass through the

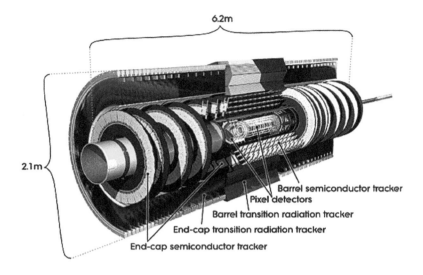

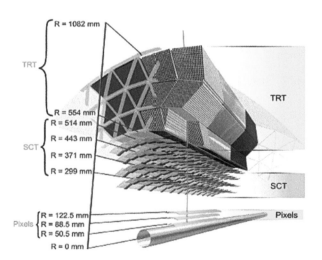

Fig. 4.6 Illustrations of the ATLAS inner detector system [4]. *Top*: overview of the ATLAS Inner Detector, with labels and dimensions. *Bottom*: drawing showing the sensors and structural elements of the ATLAS inner detector barrel being crossed by one high energy particle

TRT they emit transition radiation (TR) photons. The probability of emitting TR photons depends on the Lorentz factor, γ. Thus for particles of a given momentum, electrons will typically emit more photons than charged hadrons. Hence, the number of TR photons detected can be used to discriminate between electrons and charged hadrons.

4.2.2 Calorimeters

After particles pass through the tracking detector, they reach the calorimeters. Calorimeters are designed to provide accurate energy measurements of passing particles which undergo either electromagnetic or hadronic interactions. Only neutrinos, which don't interact in the detector, and muons, which are minimally ionizing particles (MIPs) at the LHC energy scale, travel through the calorimeters with little energy deposit. As particle(s) from the hard-scatter collision interact with the calorimeter material, they create cascades of particles called showers. Depending on the nature of the source particle, there are two types of cascades, namely electromagnetic and hadronic showers.

These two types of showers have radically different properties, and require separate techniques for high-precision detection. The electromagnetic calorimeter system is designed to measure the energy of electromagnetic showers from electrons[1] and photons. The hadronic calorimeter system is designed to measure the energy of hadrons, which are produced in collisions through quark and gluon hadronization. Both systems are classified as non-compensating sampling calorimeters. Incident particles shower in a dense passive material (e.g., lead or iron), and the shower is sampled as it passes through an active detection medium. Detector sensors only recover a fraction of the energy in the shower, but the full energy can be inferred from the observed energy. The ATLAS calorimeter system consists of both electromagnetic and hadronic calorimeters, with a total coverage of $|\eta| < 4.9$, as shown in Fig. 4.7.

The liquid-argon (LAr) electromagnetic (EM) calorimeter sits directly outside the ID and is designed to detect electrons and photons. The calorimeter uses an accordion-like structure of lead as the passive medium. The active medium, LAr, is ionized by charged particles from incident showers and the electrons produced by the ionization are collected on segmented electrodes. The system is divided into a barrel portion that covers the range $|\eta| < 1.475$, and two end cap portions on either side of the barrel covering $1.375 < |\eta| < 3.2$. It is also segmented into three layers with different $\eta-\phi$ granularities. The innermost layer has the finest η segmentation, with an η granularity of 0.003 in the barrel. The second layer has an $\eta-\phi$ granularity of 0.025×0.025 in the barrel, and is responsible for most of the energy measurement. The third layer is coarsely segmented in η but it gives the calorimeter extra depth.

The surrounding calorimeter system known as the hadronic calorimeter is responsible for measuring hadrons that either do not shower in the electromagnetic calorimeter, or the showering of which is not fully contained in the electromagnetic calorimeter. It contains barrel and end cap portions. The barrel portion, called the tile calorimeter, has coverage for $|\eta| < 1.7$. It contains alternating layers of iron as the passive medium and scintillating plastic as the active medium. Each of the cells in the tile calorimeter has an $\eta - \phi$ granularity of roughly 0.1×0.1. The end

---

[1]The electromagnetic calorimeter measures positrons as well as electrons since their showers are almost exactly the same. Positrons are not mentioned explicitly in this section to simplify the text.

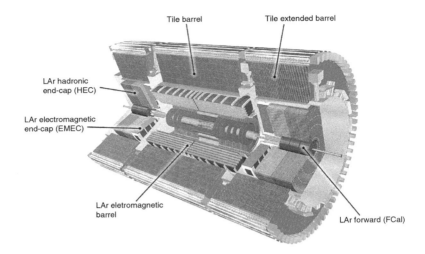

Fig. 4.7 An overview of the ATLAS calorimeter system [4]

cap portions of the calorimeter cover the region $1.5 < |\eta| < 3.2$. They use LAr technology similar to the LAr EM calorimeter, but have copper rather than lead as the passive material.

Finally, the LAr forward calorimeters are designed to measure both electromagnetic and hadronic showers. They extend the coverage to $|\eta| = 4.95$.

4.2.3 Muon Spectrometers

For particles to make it through the full depth of the calorimeters, they must either be noninteracting, a semi-stable MIP, or a shower that was not contained in the available material. The latter process is rare, though it can be important at high energies. In the first case, particles such as neutrinos, or DM particles if they exist, would simply elude the detector and present as missing energy. At typical LHC energies, the only SM example of the second case is the muon. Muons thus provide very clean signatures for both searches and measurements, and a set of detectors designed to identify and measure them has been developed.

The muon spectrometer (MS) surrounds the calorimeter systems and is the outermost detector system of ATLAS. It is designed to measure the trajectories and momenta of muons. The system contains position sensitive chambers that operate in a 0.5 T toroidal magnetic field that causes charged particles to bend in the $r - z$ plane.

An illustration of the detecting chambers and magnets is shown in Fig. 4.8. There is a large barrel toroid, which produces a field that is strongest in the range $|\eta| < 1.4$. There are also smaller toroid magnets in each end cap that produce fields strongest

4.3 Trigger and Data Acquisition System

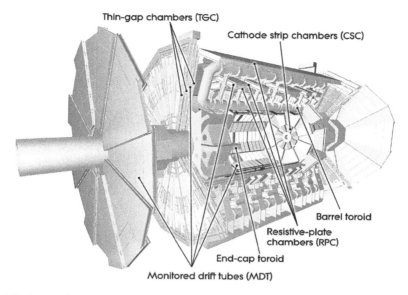

Fig. 4.8 An overview of the ATLAS muon spectrometers [4]

in the range $1.6 < |\eta| < 2.7$. In the range $1.4 < |\eta| < 1.6$, muons are bent by a combination of the barrel and end cap fields. The detector chambers are arranged into three layers in the barrel, end cap, and transition regions. In the barrel the layers are wrapped cylindrically around the beam axis. In the end caps and transition regions the layers are arranged perpendicular to the beam line.

The MS detector chambers utilize four technologies that each operate by collecting electrons that are produced when incident muons pass through a gaseous mixture. There are resistive plate chambers [20] (RPC) in the barrel that collect charge on parallel resistive plates separated by a small gap. Similarly, the end cap contains thin gap chambers [21] (TGC) consisting of two conducting cathodes separated by a gap, and two parallel wires that collect ionized charge. Both TGC and RPC provide very fast signals for muon hits, and can therefore be used in the ATLAS trigger system to select events containing muons. There are also Muon Drift Tubes [22], which are used over most of the η range, and Cathode Strip Chambers [23], which are used in the range $2 < |\eta| < 2.7$. These provide track measurements with high spatial resolution.

4.3 Trigger and Data Acquisition System

The trigger system plays an important role at a hadron collider. As many of the more interesting physics processes suffer from small cross-sections, large numbers of collisions are needed to produce sufficient quantities of these rare events. The

LHC bunch crossing rate as delivered to the ATLAS experiment for the 2012 dataset was every 50 ns, in other words a collision frequency of 20 MHz. It is both technologically impossible and ineffective for addressing the important physics questions for all the collision information from the full ATLAS detector to be read out and recorded at this rate, hence it is critical to implement a trigger system to select only the most interesting events.

4.3.1 The ATLAS Trigger System

The ATLAS trigger system is comprised of three levels, namely Level 1 (L1), Level 2 (L2), and Event Filter (EF). The system reduces the event rate by a factor of 10^5 from 20 MHz to 200 Hz.

The L1 trigger selection is based on basic energy clustering in the calorimeter towers (geometrical groups of calorimeter cells) for electron/photon and jet reconstruction, or in the case of muons, track finding in trigger-specific chambers the muon system, as these are the only parts of the detector which can perform a simple reconstruction and trigger decision within 2.5 µs on an event-by-event basis. The L1 trigger analyzes all 20 million events per second, and selects one per $\sim$500 events to proceed to the next level. To fulfill the processing speed necessary, the L1 trigger reconstruction and selection algorithms are implemented directly in hardware. If the L1 trigger decides an event was interesting, it defines one or more regions of interest (RoIs) specifying the η and ϕ coordinates of the interesting object(s), and the trigger requirement that were fulfilled.

This information is then sent to the software-based high-level trigger (HLT), which includes both the L2 trigger and the EF. The L2 trigger matches inner detector information to the RoIs found by the L1 trigger, and decides whether the event still passes the appropriate trigger criteria within 40 ms, reducing the event rate from 50 kHz to 5 kHz. Successful events are then passed to the EF, which further reduces the trigger rate to the final level of 300 Hz. This is done by performing a full event reconstruction similar to what is used for offline analysis, including calibrations, alignment corrections, and advanced algorithms. If an event is marked as interesting by the EF, it is recorded to disk in the RAW data format, which includes the information on all of the digitized energy deposits throughout the entire ATLAS detector.

For the analysis presented in this thesis, events used are triggered by the reconstructed missing transverse momentum, with additional usage of events triggered by reconstructed muons, electrons, photons, and jets. The details of the reconstructed physics objects and the trigger selection are described in Chap. 8.

4.3.2 Fast Tracker Track Trigger Upgrade

The existing ATLAS trigger system was designed to work well at the LHC design luminosity. However, after the planned luminosity upgrade, the increase in detector activity arising from many simultaneous interactions, and an increased rate to the HLT from upgrades to the L1 trigger, will make extensive tracking, critical to event selection and physics analysis, prohibitively expensive in terms of processing time per event or computing cores needed. The existing approach of performing tracking for specific RoIs already identified by the L1 trigger and full event tracking at low rates of a few kHz has several limitations. First, there is a limit to either the number or size of RoIs processed by the HLT, which forces additional nontracking cuts to be applied, resulting in reduced efficiency or higher thresholds for the objects considered. Second, there are cases where global event information, such as the location of the hard interaction vertex or number of primary vertices in the event, are useful for object selections or corrections to the other detector quantities.

In light of these limitations, and motivated by the importance of good b-quark and τ lepton selection especially important for physics analysis at LHC RunII and beyond, a hardware-based trigger upgrade system is being constructed, called the Fast Tracker or FTK, which performs global track reconstruction after each L1 trigger to grant the L2 trigger early access to tracking information. FTK utilizes state-of-the-art electronics and a highly parallel processing system to rapidly find and reconstruct tracks in the inner detector pixel and SCT layers for every event that passes the level-1 trigger, over the full rapidity range covering both the barrel and the end cap regions. FTK receives the hits at full rate as they are sent from the pixel and SCT detector readout drivers following a L1 trigger. After processing, FTK fills the readout system buffers with the helix parameters and hits for all tracks with p_T above a minimum value, typically 1 GeV. The L2 processors then have the option of requesting the track information in either a RoI or the entire detector. Figure 4.9 illustrates the layout of the ATLAS trigger system, and the positioning of the FTK within it.

FTK is based on the very successful silicon vertex trigger [24] used for the CDF experiment at the Tevatron. Extensive physics performance studies in simulation show that the usage of FTK will bring significant improvements to various physics objectives in both precision measurements of known processes, or discovery of new physics. Among these objectives are the identification of tracks and the primary vertices, to the reconstruction and identification of more complicated physics objects like b-quark jets, τs, and other leptons, and potential new particles that display a few tracks in a narrow cone or a displaced vertex when passing through a detector. More details on the design and performance of FTK are documented in its Technical Design Report [25].

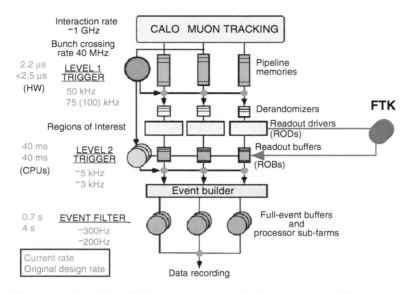

Fig. 4.9 An overview of the ATLAS trigger system and its integration with FTK

References

1. T.S. Pettersson, P Lefèvre, The Large Hadron Collider: conceptual design. Technical Report CERN-AC-95-05-LHC (CERN, Geneva, 1995)
2. L. Evans, P. Bryant, LHC machine. J. Instrum. **3**, S08001 (2008)
3. O.S. Bruning, P. Collier, P. Lebrun, S. Myers, R. Ostojic, J. Poole, P. Proudlock, *LHC Design Report* (CERN, Geneva, 2004)
4. ATLAS Collaboration, The ATLAS experiment at the CERN Large Hadron Collider. J. Instrum. **3**, S08003 (2008)
5. ATLAS Collaboration, ATLAS: detector and physics performance technical design report. Volume 1. ATLAS-TDR-14 (1999)
6. ATLAS Collaboration, Studies of the performance of the ATLAS detector using cosmic-ray muons. Eur. Phys. J. C **71**, 1593 (2011)
7. ATLAS Collaboration, The ATLAS inner detector commissioning and calibration. Eur. Phys. J. C **70**, 787–821 (2010)
8. ATLAS Collaboration, Performance of the ATLAS detector using first collision data. J. High Energy Phys. **1009**, 056 (2010)
9. S. Chatrchyan et al., The CMS experiment at the CERN LHC. J. Instrum. **3**, S08004 (2008)
10. K. Aamodt et al., The ALICE experiment at the CERN LHC. J. Instrum. **3**, S08002 (2008)
11. Jr. Alves, A. Augusto et al., The LHCb detector at the LHC. J. Instrum. **3**, S08005 (2008)
12. LHC Programme Coordination, Online. https://lpc.web.cern.ch/lpc/lumiplots_2012.htm
13. ATLAS Collaboration, Online. https://twiki.cern.ch/twiki/bin/view/AtlasPublic/LuminosityPublicResults
14. ATLAS Collaboration, Luminosity determination in pp collisions at $\sqrt{s} = 7$ TeV using the ATLAS detector at the LHC. Eur. Phys. J. C **71**, 1630 (2011)
15. ATLAS Collaboration, ATLAS pixel detector: technical design report. CERN-LHCC-98-13 (1998)

References

16. G. Aad, M. Ackers, F.A. Alberti, M. Aleppo, G. Alimonti et al., ATLAS pixel detector electronics and sensors. J. Instrum. **3**, P07007 (2008)
17. A. Abdesselam, T. Akimoto, P.P. Allport, J. Alonso, B. Anderson et al., The barrel modules of the ATLAS semiconductor tracker. Nucl. Instrum. Methods A **568**, 642–671 (2006)
18. A. Abdesselam et al., The ATLAS semiconductor tracker end-cap module. Nucl. Instrum. Meth. A **575**, 353–389 (2007)
19. E. Abat et al., The ATLAS transition radiation tracker (TRT) proportional drift tube: design and performance. J. Instrum. **3**, P02013 (2008)
20. G. Aielli, A. Aloisio, M. Alviggi, V. Aprodu, V. Bocci et al., The RPC first level muon trigger in the barrel of the ATLAS experiment. Nucl. Phys. Proc. Suppl. **158**, 11–15 (2006)
21. S. Majewski, G. Charpak, A. Breskin, G. Mikenberg, A thin multiwire chamber operating in the high multiplication mode. Nucl. Instrum. Methods **217**, 265–271 (1983)
22. F. Bauer, U. Bratzler, H. Dietl, H. Kroha, T. Lagouri et al., Construction and test of MDT chambers for the ATLAS muon spectrometer. Nucl. Instrum. Methods A **461**, 17–20 (2001)
23. T. Argyropoulos, K.A. Assamagan, B.H. Benedict, V. Chernyatin, E. Cheu et al., Cathode strip chambers in ATLAS: installation, commissioning and in situ performance. IEEE Trans. Nucl. Sci. **56**, 1568–1574 (2009)
24. S. Belforte et al., The CDF trigger silicon vertex tracker (SVT). IEEE Trans. Nucl. Sci. **42**, 860–864 (1995)
25. M. Shochet, L. Tompkins, V. Cavaliere, P. Giannetti, A. Annovi, G. Volpi, Fast TracKer (FTK) technical design report. Technical Report CERN-LHCC-2013-007. ATLAS-TDR-021 (CERN, Geneva, 2013). ATLAS Fast Tracker technical design report

Chapter 5
Dark Matter Searches at ATLAS

This chapter gives an overview of searches for dark matter (DM) at the ATLAS experiment of the Large Hadron Collider (LHC). A generalized introduction to DM is given in Chap. 3, in which Sect. 3.5 presents the three complementary approaches of DM searches, namely direct detection, indirect detection, and collider production. Similar collider-based searches for DM have been conducted by the CMS experiment at the LHC as well, and to a lesser extent at the Tevatron, but they are not discussed here.

This chapter is organized as follows: Sect. 5.1 describes the theoretical basis for collider searches for DM, including both DM produced through a contact operator in the effective field theory framework (Sect. 5.1.1) and simplified models of DM-SM interactions; based on the production mechanism and hence the collider signature, a series of searches for DM at ATLAS conducted during RunI of the LHC are listed in Sect. 5.2; one such search, where DM is produced in association with heavy quarks, i.e., bottom or top quarks, is described in more detail in Sect. 5.3, as it leads to a similar final state and served as a predecessor for the DM + Higgs($\to b\bar{b}$) search which is the focus of this thesis.

5.1 Theoretical Models

While the existence of DM is well established, as discussed in Sect. 3.3, very little is known about the properties of the DM particle(s) and how they may interact with SM particles. Hence, one can construct a large number of qualitatively different DM models, populating the "theory space" of all possible realizations of physics beyond the SM with a particle that is a viable DM candidate. For the interest of existing collider searches, the DM particles in discussion are typically assumed to be WIMPs. As illustrated in Fig. 5.1, the members of this theory space fall into three

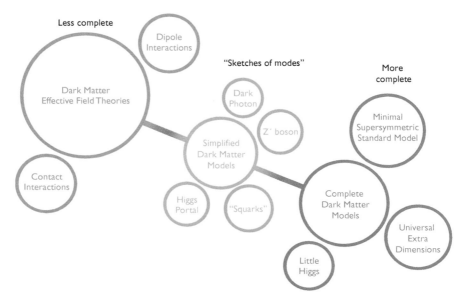

Fig. 5.1 Overview of the DM theory space, from left to right by order of ascending degree of completeness, EFT approach, simplified models, and UV-complete models [1]

distinct classes based on the completeness of the models: the effective field theory (EFT) approach, simplified models, and ultra-violet (UV) complete models.

The last one, UV-complete models, have been discussed briefly in Sect. 3.4 as extensions to the SM, like SUSY or extra-dimensions, and may contain particles that make viable DM candidates. DM particles predicted by these models can be searched for directly at collider experiments as part of a generalized search for these BSM theories, and are the most model-dependent. In other words, their structures are so rich that it may be very difficult to determine unambiguously the underlying new dynamics from a finite amount of data ("inverse problem"). They are not further discussed here. The following paragraphs present an overview of the EFT approach and simplified models of DM production at colliders.

5.1.1 Effective Field Theory Framework

On the simple end of the spectrum, EFT allows one to describe the DM-SM interactions mediated by a heavy particle kinematically inaccessible at collider experiments, and hence it is treated as a contact operator in a model-independent approach. EFTs are very powerful tools, as they allow for the reduction of a complicated process to the minimal number of degrees of freedom. The variables of interest are reduced to two parameters per mediator type, namely the mass of the DM particle (m_χ), and the suppression scale Λ or the coupling strength

5.1 Theoretical Models

Table 5.1 Operators coupling WIMPs to SM particles [2]

Name	Operator	Coefficient
D1	$\bar{\chi}\chi\bar{q}q$	m_q/M_*^3
D2	$\bar{\chi}\gamma^5\chi\bar{q}q$	im_q/M_*^3
D3	$\bar{\chi}\chi\bar{q}\gamma^5 q$	im_q/M_*^3
D4	$\bar{\chi}\gamma^5\chi\bar{q}\gamma^5 q$	m_q/M_*^3
D5	$\bar{\chi}\gamma^\mu\chi\bar{q}\gamma_\mu q$	$1/M_*^2$
D6	$\bar{\chi}\gamma^\mu\gamma^5\chi\bar{q}\gamma_\mu q$	$1/M_*^2$
D7	$\bar{\chi}\gamma^\mu\chi\bar{q}\gamma_\mu\gamma^5 q$	$1/M_*^2$
D8	$\bar{\chi}\gamma^\mu\gamma^5\chi\bar{q}\gamma_\mu\gamma^5 q$	$1/M_*^2$
D9	$\bar{\chi}\sigma^{\mu\nu}\chi\bar{q}\sigma_{\mu\nu}q$	$1/M_*^2$
D10	$\bar{\chi}\sigma_{\mu\nu}\gamma^5\chi\bar{q}\sigma_{\alpha\beta}q$	i/M_*^2
D11	$\bar{\chi}\chi G_{\mu\nu}G^{\mu\nu}$	$\alpha_s/4M_*^3$
D12	$\bar{\chi}\gamma^5\chi G_{\mu\nu}G^{\mu\nu}$	$i\alpha_s/4M_*^3$
D13	$\bar{\chi}\chi G_{\mu\nu}\tilde{G}^{\mu\nu}$	$i\alpha_s/4M_*^3$
D14	$\bar{\chi}\gamma^5\chi G_{\mu\nu}\tilde{G}^{\mu\nu}$	$\alpha_s/4M_*^3$

Name	Operator	Coefficient
C1	$\chi^\dagger\chi\bar{q}q$	m_q/M_*^2
C2	$\chi^\dagger\chi\bar{q}\gamma^5 q$	im_q/M_*^2
C3	$\chi^\dagger\partial_\mu\chi\bar{q}\gamma^\mu q$	$1/M_*^2$
C4	$\chi^\dagger\partial_\mu\chi\bar{q}\gamma^\mu\gamma^5 q$	$1/M_*^2$
C5	$\chi^\dagger\chi G_{\mu\nu}G^{\mu\nu}$	$\alpha_s/4M_*^2$
C6	$\chi^\dagger\chi G_{\mu\nu}\tilde{G}^{\mu\nu}$	$i\alpha_s/4M_*^2$

The operator names beginning with D and C apply to WIMPS that are Dirac fermions and complex scalars, respectively

λ of the interaction. One could write down the Lagrangian of the operators at various dimensions, which describe the interaction between DM and SM particles. Following the convention in [2], considering all possible couplings via spin 0, 1, and 2 mediators between partons and both fermionic and scalar DM, the operators are listed in Table 5.1. Here, the suppression scale is expressed as the effective mass of the mediator, denoted as M_*.

The 20 types of operators in Table 5.1 can be further simplified by grouping together the ones with similar kinematic properties at collider searches. For example, while the chirality of the operators affect direct and indirect detection results, for collider searches it only manifests as changes in cross-section. In the end, the 14 fermionic operators can be broken down to four families, and the six complex scalar operators into three, of which one fermionic operator and one scalar operator have the same kinematic distributions, yielding six distinct operators in total, namely D1, C1, D5 (C3), D9, D11, and C5. This categorization allows for a full coverage of a large parameter space with a minimal set of simulated samples.

While the aforementioned EFT operators have served as benchmark models for LHC searches for DM, especially in cases when there is a jet produced recoiling against the missing energy from the DM particles, an abundance of other EFT operators have also been proposed, some of which yielding associated production of specific visible particle(s) like a photon, a W or Z boson, or a Higgs boson, etc.

The EFT approach is very useful as it can yield stringent bounds on the "new-physics" scale M_* that suppresses the higher-dimensional operators. Since for each operator, a single parameter contains the information on all the heavy states of the dark sector, comparing LHC bounds to the limits from direct and indirect DM searches is straightforward in the EFT framework.

However, the EFT approach is based on the assumption that the mediator particle is too heavy to be produced directly and hence can be integrated out of the EFT framework and the process treated as a contact interaction. This is not valid when the momentum transferred in the interaction, Q_{tr}, is comparable to the mass of the mediator [3, 4]. To address this issue and give an indication of the impact of the unknown ultraviolet details of the theory, a truncation method [5] is applied to naive collider limits on EFT operators, where only simulated events with $Q_{tr} =< M_*$ are retained with bounds on the effective coupling to be below 4π, which is the maximum possible value for the interaction to remain perturbative.

5.1.2 Simplified Models

As the EFT approach becomes a poor approximation of the UV-complete models when the momentum transfer is large, as it can be at the energies accessible at the LHC, it is important to expand the level of detail with regard to DM-SM interactions, and hence develop the so-called simplified models (e.g., [1]). Such models are characterized by the most important state mediating the DM particle interactions with the SM and hence explicitly include the particles at higher masses, as well as the DM particle itself. Unlike the EFT approach, simplified models are able to describe correctly the full kinematics of DM production at the LHC, because they resolve the EFT contact interactions into single-particle s-channel or t-channel exchanges. This comes with the price that they typically involve not just one, but a handful of parameters that characterize the dark sector and its coupling to the visible sector. The increased number of assumptions makes such models less generic.

For a simplified DM model to be useful at the LHC, it should be simple enough to contain only the minimal amount of renormalizable interactions as opposed to a more complicated model, and be complete enough to describe accurately the relevant physics phenomena at the energies that can be probed at the LHC. Generally speaking, the following requirements on the particle content and the interactions of the simplified model should be met:

1. Besides the SM, the model should contain a DM candidate that is stable enough to escape the LHC detectors, as well as a mediator that couples the two sectors. The dark sector can be richer, but the additional states should be somewhat decoupled.
2. The Lagrangian should in principle contain all terms that are renormalizable and consistent with Lorentz invariance, the SM gauge symmetries, and DM stability.
3. The additional interactions should not violate the exact and approximate accidental global symmetries of the SM. This means that the interactions between the visible and the dark sector should be such that baryon and lepton number are conserved and that the custodial and flavor symmetries of the SM are not strongly broken.

A variety of simplified models for DM production have been proposed in recent years, covering a wide range of cases of spin-0 or spin-1, s-channel or t-channel exchanges, Higgs-portal, etc. An overview of the most relevant simplified models for LHC searches are given in [6].

5.2 "Mono-jet" and other "Mono-X" Searches

To search for DM production at collider experiments, the minimal experimental signature consists of an excess of events with one or more visible final-state object, X, recoiling against large amounts of missing transverse momentum (E_T^{miss}), where the E_T^{miss} may be interpreted as from DM particles that were produced and escaped the detectors. In Run I of the LHC, the ATLAS and CMS collaborations have examined a variety of such "mono-X" signatures, considering "X" to be a hadronic jet [7, 8], heavy-flavor jet [9, 10], photon [11, 12], or W/Z boson [13, 14].

While certain "mono-X" signatures arise from the production of "partner" particles that decay to DM and Standard Model (SM) particles (which usually leads to more complex final states), for many existing "mono-X" searches for DM, the visible signature of "X" arises from initial (final) state radiation (ISR or FSR) of an incoming (outgoing) quark or gluon, regardless of the specifics of the mediating particle from which the DM particles are pair produced. This means while some models only allow a specific "X" associated production, searches for a variety of "mono-X" final states can often yield comparable results on the same set of models, one of which may produce a stronger bound on the specific model depending on the underlying theory that describes the interaction. Generally speaking, the most stringent bounds on a number of benchmark DM models are found from "mono-jet" searches, which is the search for events involving a single jet balanced against large E_T^{miss}. Figure 5.2 shows two example Feynman diagrams of DM production with ISR jet emission for either a quark EFT operator (left) or a Z' mediator in a simplified model (right).

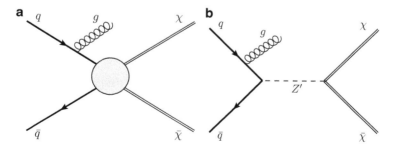

Fig. 5.2 Feynman diagrams for the production of WIMP pairs associated with a jet from initial-state radiation of a gluon, g. (**a**) A contact interaction described with effective operators. (**b**) A simplified model with a Z' boson [7]

The discovery of the Higgs boson h [15, 16] provides a new opportunity to search for DM production via the $h + E_T^{miss}$ signature [17, 18]. The first collider search for DM production in association with a Higgs boson was conducted at ATLAS using the Higgs decay channel of $h \to \gamma\gamma$ [19]. The focus of this thesis, concerning the Higgs decay into two bottom quarks, will be discussed in later chapters.

5.3 Dark Matter Search with Heavy Flavor

In a typical "mono-jet" search for DM, the jet is predominantly a light-quark jet or a gluon jet. A new search for DM pair production in association with one b-quark or a pair of heavy quarks (b or t) was proposed in [20].

Searches for such final states are particularly sensitive to effective scalar interactions between DM and quarks described by the operator (Table 5.1, [3])

$$\mathcal{O}_{\text{scalar}} = \sum_q \frac{m_q}{M_*^N} \bar{q}q\bar{\chi}\chi, \qquad (5.1)$$

where $N = 3$ for Dirac DM (D1 operator) and $N = 2$ for complex scalar DM (C1 operator), and the quark and DM fields are denoted by q and χ, respectively. The scalar operators are normalized by m_q, which mitigates contributions to flavor-changing processes through the framework of minimal flavor violation (MFV). The dependence on the quark mass makes final states with heavy, i.e., bottom and top quarks the most sensitive to these operators.

Additionally, the tensor operator (D9), which describes a magnetic moment coupling, is parameterized as (Table 5.1, [3]) :

$$\mathcal{O}_{\text{tensor}} = \sum_q \frac{1}{M_*^2} \bar{\chi}\sigma^{\mu\nu}\chi \bar{q}\sigma_{\mu\nu}q. \qquad (5.2)$$

MFV suggests that the D9 operator should have a mass dependence from Yukawa couplings, making this search sensitive to tensor couplings as well.

The dominant Feynman diagrams for these processes are shown in Fig. 5.3, with final states characterized by the presence of jets originating from b-quarks, missing transverse momentum, and leptons (e, μ) in the case of a semileptonic decay of a top quark.

Moreover, to explain the excess of gamma rays from the galactic center, recently observed by the Fermi Gamma-ray Space Telescope, and interpreted as a signal for DM annihilation [21], a bottom-flavored dark matter model (b-FDM) [22] was proposed as shown in Fig. 5.4. The collider signature of this model is b-quarks produced in association with missing transverse momentum.

Details of this search, including its results interpreted as constraints on the mass scales of the EFT operators and exclusion of regions of parameter space for

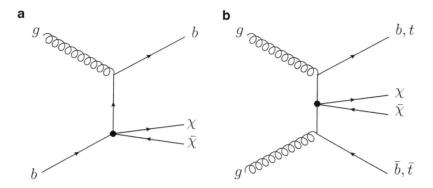

Fig. 5.3 Dominant Feynman diagrams for DM production in conjunction with (**a**) a single *b*-quark and (**b**) a heavy quark (*bottom* or *top*) pair using an EFT approach

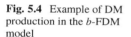

Fig. 5.4 Example of DM production in the *b*-FDM model

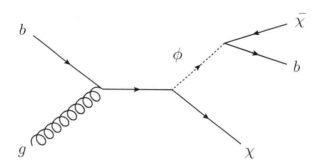

the *b*-FDM, can be found in [9]. The exclusion limits are strongest at low DM masses. The limit on the χ–nucleon cross-section mediated by the D1 operator is improved significantly with respect to previously published ATLAS results by obtaining sensitivities of approximately $\sigma^{SI}_{\chi-N} = 10^{-42}$ cm^2 for $m_\chi = 10$ GeV.

References

1. J. Abdallah et al., Simplified models for dark matter searches at the LHC. Phys. Dark Univ. **9–10**, 8–23 (2015)
2. J. Goodman, M. Ibe, A. Rajaraman, W. Shepherd, M.P.T. Tait, H. Yu, Constraints on dark matter from colliders. Phys. Rev. D **82**, 116010 (2010)
3. J. Goodman, M. Ibe, A. Rajaraman, W. Shepherd, T.M.P. Tait, H.-B. Yu, Constraints on dark matter from colliders. Phys. Rev. **D82**, 116010 (2010)
4. G. Busoni, A. De Simone, E. Morgante, A. Riotto, On the validity of the effective field theory for dark matter searches at the LHC. Phys. Lett. B **728**, 412–421 (2014)
5. G. Busoni, A. De Simone, J. Gramling, E. Morgante, A. Riotto, On the validity of the effective field theory for dark matter searches at the LHC, part II: complete analysis for the *s*-channel. J. Cosmol. Astropart. Phys. **1406**, 060 (2014)

6. D. Abercrombie et al., Dark matter benchmark models for early LHC run-2 searches: report of the ATLAS/CMS dark matter forum. Submitted to Phys. Dark Univ. arXiv:1507.00966 [hep-ex] (2015)
7. ATLAS Collaboration, Search for new phenomena in final states with an energetic jet and large missing transverse momentum in pp collisions at $\sqrt{s} = 8$ TeV with the ATLAS detector. Eur. Phys. J. C **75**, 299 (2015)
8. CMS Collaboration, Search for dark matter, extra dimensions, and unparticles in monojet events in proton–proton collisions at $\sqrt{s} = 8$ TeV. Eur. Phys. J. C **75**, 235 (2015)
9. ATLAS Collaboration, Search for dark matter in events with heavy quarks and missing transverse momentum in *pp* collisions with the ATLAS detector. Eur. Phys. J. C **75**, 92 (2015)
10. CMS Collaboration, Search for monotop signatures in proton-proton collisions at $\sqrt{s} = 8$ TeV. Phys. Rev. Lett. **114**(10), 101801 (2015)
11. ATLAS Collaboration, Search for new phenomena in events with a photon and missing transverse momentum in *pp* collisions at $\sqrt{s} = 8$ TeV with the ATLAS detector. Phys. Rev. D **91**, 012008 (2015)
12. CMS Collaboration, Search for dark matter and large extra dimensions in pp collisions yielding a photon and missing transverse energy. Phys. Rev. Lett. **108**, 261803 (2012)
13. ATLAS Collaboration, Search for dark matter in events with a Z boson and missing transverse momentum in pp collisions at $\sqrt{s} = 8$ TeV with the ATLAS detector. Phys. Rev. D **90**, 012004 (2014)
14. ATLAS Collaboration, Search for dark matter in events with a hadronically decaying W or Z boson and missing transverse momentum in pp collisions at $\sqrt{s} = 8$ TeV with the ATLAS detector. Phys. Rev. Lett. **112**, 041802 (2014)
15. ATLAS Collaboration, Observation of a new particle in the search for the Standard Model Higgs boson with the ATLAS detector at the LHC. Phys. Lett. B **716**, 1–29 (2012)
16. CMS Collaboration, Observation of a new boson at a mass of 125 GeV with the CMS experiment at the LHC. Phys. Lett. B **716**, 30–61 (2012)
17. L. Carpenter et al., Mono-Higgs-boson: a new collider probe of dark matter. Phys. Rev. D **89**, 075017 (2014)
18. A. Berlin, T. Lin, L.-T. Wang, Mono-Higgs detection of dark matter at the LHC. J. High Energy Phys. **06**, 078 (2014)
19. ATLAS Collaboration, Search for dark matter in events with missing transverse momentum and a Higgs boson decaying to two photons in *pp* collisions at $\sqrt{s} = 8$ TeV with the ATLAS detector. Phys. Rev. Lett. **115**(13), 131801 (2015)
20. T. Lin, E.W. Kolb, L.-T. Wang, Probing dark matter couplings to top and bottom quarks at the LHC. Phys. Rev. D **88**(6), 063510 (2013)
21. T. Daylan, D.P. Finkbeiner, D. Hooper, T. Linden, S.K.N. Portillo et al., The characterization of the gamma-ray signal from the central milky way: a compelling case for annihilating dark matter. (2014)
22. P. Agrawal, B. Batell, D. Hooper, T. Lin, Flavored dark matter and the galactic center gamma-ray excess. Phys. Rev. D **90**(6), 063512 (2014)

Chapter 6
Dark Matter + Higgs($\to b\bar{b}$): Overview

This chapter presents an overview of the analysis in search of dark matter (DM) produced in association with a Higgs boson in the $h \to b\bar{b}$ channel [1]. For other types of collider searches for DM, see Chap. 5. This chapter is organized as follows: Sect. 6.1 briefly states the physics motivation for such a search; the models used for this analysis, including both EFT models and a simplified model, are introduced in Sect. 6.2; and Sect. 6.3 describes the two analysis methods used in this analysis, categorized by the way the Higgs boson is reconstructed.

6.1 Physics Motivation

The discovery of the Higgs boson h [2, 3] provides a new opportunity to search for DM production via the $h + E_T^{miss}$ signature [4–6]. In contrast to most of the aforementioned probes, the visible Higgs boson is unlikely to have been radiated from an initial-state quark or gluon, and the signal would give insight into the structure of DM coupling to SM particles.

The decay channel with the largest branching ratio for the discovered 125 GeV Higgs boson is the $h \to b\bar{b}$ channel. Preliminary sensitivity studies showed a search in the $b\bar{b}$ final state would yield significantly stronger limits compared with an earlier ATLAS search in the $h \to \gamma\gamma$ channel [6].

Moreover, unlike previous ATLAS searches for resonant production with a similar final state [7, 8], this analysis explores different theoretical models, focuses on the fully hadronic channel with data-driven methods to estimate the main backgrounds, and most importantly, applies selections extending to large E_T^{miss} utilizing "resolved" as well as "boosted" techniques to be described later. The approach for extracting limits in this analysis is also more suited for the models considered here, and reduces the theoretical uncertainty from modeling and fitting of the signal shape.

All these reasons make a compelling case to perform such an analysis.

6.2 Signal Models

As described in Sect. 5.1, two approaches are commonly used to model generic processes yielding a final state with a particle X recoiling against a system of noninteracting particles, either with an effective field theory (EFT) framework [9], where particles that mediate the interactions between DM and SM particles are too heavy to be produced directly in the experiment and are described by contact operators, or using simplified models that are characterized by a minimal number of renormalizable interactions and hence explicitly include the particles at higher masses [10]. The two approaches are complementary and both are included in this analysis.

6.2.1 EFT Models

Using the EFT approach, a set of models described by effective operators at different dimensions is considered. Following the notation in [4], the effective operators in ascending order of their dimensions are:

$$\lambda |\chi|^2 |H|^2 \quad \text{(Scalar DM, dimension-4)} \quad (6.1)$$

$$\frac{1}{\Lambda} \bar{\chi} i \gamma_5 \chi |H|^2 \quad \text{(Fermionic DM, dimension-5)} \quad (6.2)$$

$$\frac{1}{\Lambda^2} \chi^\dagger \partial^\mu \chi H^\dagger D_\mu H \quad \text{(Scalar DM, dimension-6)} \quad (6.3)$$

$$\frac{1}{\Lambda^4} \bar{\chi} \gamma^\mu \chi B_{\mu\nu} H^\dagger D^\nu H \quad \text{(Fermionic DM, dimension-8)} \quad (6.4)$$

Here χ is the DM particle, which is a gauge singlet under $SU(3)_C \times SU(2)_L \times U(1)_Y$ and may be a scalar or a fermion as specified, $D_\mu(^\nu)$ is the covariant derivative for the full gauge group, and $B_{\mu\nu}$ is the $U(1)_Y$ field strength tensor. The parameters of these models are the DM particle mass m_χ, and the coupling parameter λ or the suppression scale Λ of the heavy mediator that is not directly produced but is described by a contact operator in the EFT framework.

6.2.2 Simplified Models

A simplified model is also considered which contains a Z' gauge boson and two Higgs fields resulting in five Higgs bosons (often called the two-Higgs-doublet model, 2HDM) [5], where the DM particle is coupled to the heavy pseudoscalar Higgs boson A. This model, named Z'-2HDM, is described in detail in Sect. 7.

Other types of simplified models yielding a DM plus Higgs boson signature have been proposed as well, such as a baryonic or "Hidden Valley" Z' mediator or a scalar mediator [4]: they are not used in this analysis and hence not discussed here.

6.3 Analysis Channels

Two Higgs boson reconstruction techniques are presented that are complementary in their acceptance, leading to two analysis channels, the "resolved" channel where the Higgs boson is reconstructed as two separate b-jets, and the "boosted" channel where the Higgs boson is reconstructed as a single large-radius jet.

6.3.1 Resolved Analysis

The first "resolved" technique reconstructs Higgs boson candidates from pairs of nearby anti $- k_\mathrm{T}$ jets [11] each reconstructed with radius parameter $R = 0.4$, each identified as having a b-hadron within the jet using a multivariate b-tagging algorithm [12]. Details of the reconstruction of the physics objects in the resolved channel is given in Chap. 8, and the signal selection described in Chap. 9. This resolved technique offers good efficiency over a wide kinematic range with the Higgs boson transverse momentum p_T between 150 and 450 GeV.

6.3.2 Boosted Analysis

For a Higgs boson with $p_\mathrm{T} \gtrsim 450\,\mathrm{GeV}$, the high momentum ("boost") of the Higgs boson causes the two jet cones containing the b- and $\bar{b}$-quarks from the Higgs boson decay to significantly overlap, leading to a decrease in the reconstruction efficiency of the two b-tagged anti $- k_\mathrm{T}$ jets with $R = 0.4$. This motivates the use of a second "boosted" Higgs boson reconstruction technique, which maintains acceptance for these higher-p_T Higgs bosons through the use of a set of jet reconstruction and identification methods that exploit the internal structure of jets, known as "jet substructure" techniques [13]. The Higgs boson candidate is reconstructed as a single anti $- k_\mathrm{T}$ $R = 1.0$ jet, trimmed with subjet radius parameter $R_\mathrm{sub} = 0.3$ and subjet transverse momentum fraction $p_{Ti}/p_T^{jet} < 0.05$, where p_{Ti} is the transverse momentum of the ith subjet and p_T^{jet} is the p_T of the untrimmed jet [14]. This $R = 1.0$ jet must be associated with two b-tagged anti $-k_T$ $R = 0.3$ jets reconstructed only from charged particle tracks (track-jets) [15]. The use of track-jets with a smaller R parameter allows the decay products of Higgs bosons with higher p_T to be reconstructed.

A set of common preselection criteria is used for events to be considered for the resolved and boosted channels. The final signal regions are defined with four increasing thresholds for the missing transverse momentum in the resolved channel, and two thresholds in the boosted channel. To search for the possible presence of non-SM signals, the total numbers of observed events after applying all selection criteria are compared with the total number of expected SM events taking into account their respective uncertainties in both channels.

The interplay between the two sets of models and analysis methods has been studied. In the Z'-2HDM simplified model, the resonant production and decay of the Z' boson leads to clear peaks in the E_T^{miss} spectra, the positions of which depend on the Z' and A mass values. In most of the parameter space probed with Z' mass between 600 and 1400 GeV, and A mass between 300 and 800 GeV (where kinematically allowed), a higher signal sensitivity is achieved in the resolved channel. On the other hand, the EFT models display very different kinematics with wide tails in high E_T^{miss} extending beyond 450 GeV, warranting a "boosted" reconstruction of the Higgs boson. Given the clear advantage of one analysis channel over the other for either set of models, and for simplicity, the results for the Z'-2HDM model are given using the resolved analysis, and the EFT models are interpreted using the boosted analysis.

The resolved channel and its interpretation are the focus of this thesis, each aspect of which will be presented in detail in the ensuing chapters.

References

1. ATLAS Collaboration, Search for dark matter produced in association with a Higgs boson decaying to two bottom quarks in pp collisions at $\sqrt{s} = 8$ TeV with the ATLAS detector. Phys. Rev. D (2015). arXiv:1510.06218 [hep-ex]
2. ATLAS Collaboration, Observation of a new particle in the search for the Standard Model Higgs boson with the ATLAS detector at the LHC. Phys. Lett. B **716**, 1–29 (2012)
3. CMS Collaboration, Observation of a new boson at a mass of 125 GeV with the CMS experiment at the LHC. Phys. Lett. B **716**, 30–61 (2012)
4. L. Carpenter et al., Mono-Higgs-boson: a new collider probe of dark matter. Phys. Rev. **D89**, 075017, 2014.
5. A. Berlin, T. Lin, L.-T. Wang, Mono-Higgs detection of dark matter at the LHC. J. High Energy Phys. **06**, 078 (2014)
6. ATLAS Collaboration, Search for dark matter in events with missing transverse momentum and a Higgs boson decaying to two photons in pp collisions at $\sqrt{s} = 8$ TeV with the ATLAS detector. Phys. Rev. Lett. **115**(13), 131801 (2015)

References

7. ATLAS Collaboration, Search for a new resonance decaying to a W or Z boson and a Higgs boson in the $\ell\ell/\ell\nu/\nu\nu + b\bar{b}$ final states with the ATLAS detector. Eur. Phys. J. **C75**(6), 263 (2015)
8. ATLAS Collaboration, Search for a CP-odd Higgs boson decaying to Zh in pp collisions at $\sqrt{s} = 8$ TeV with the ATLAS detector. Phys. Lett. **B744**, 163–183 (2015)
9. D. Abercrombie et al., Dark matter benchmark models for early LHC run-2 searches: report of the ATLAS/CMS dark matter forum. Phys. Dark Universe (2015, submitted). arXiv:1507.00966 [hep-ex]
10. J. Abdallah et al., Simplified models for dark matter searches at the LHC. Phys. Dark Universe **9–10**, 8–23 (2015)
11. M. Cacciari, G.P. Salam, G. Soyez, The anti-k_t jet clustering algorithm. J. High Energy Phys. **04**, 063 (2008)
12. ATLAS Collaboration, Calibration of the performance of b-tagging for c and light-flavour jets in the 2012 ATLAS data. (ATLAS-CONF-2014-046) (2014)
13. ATLAS Collaboration, Jet mass and substructure of inclusive jets in $\sqrt{s} = 7$ TeV pp collisions with the ATLAS experiment. J. High Energy Phys. **05**, 128 (2012)
14. D. Krohn, J. Thaler, L.-T. Wang, Jet trimming. J. High Energy Phys. **02**, 084 (2010)
15. ATLAS Collaboration, Search for Higgs boson pair production in the $b\bar{b}b\bar{b}$ final state from pp collisions at $\sqrt{s} = 8$ TeV with the ATLAS detector. Eur. Phys. J. **C75**, 412 (2015)

Chapter 7
Dark Matter + Higgs($\to b\bar{b}$): Z'-2HDM Simplified Model

This chapter gives a description of the simplified model used in the resolved analysis, (where the Higgs boson is reconstructed as two separate b-quark jets), which is the focus of this thesis. An overview of the "mono-Higgs" models for DM search is presented in Chap. 6. Results of the search for DM produced through this simplified model are presented in Chap. 12.

This chapter is organized as follows: Sect. 7.1 describes the construct of the model, the theoretical constraints on couplings and parameter space, and how the model gives rise to the "DM+Higgs" signature; Sect. 7.2 describes the regions of parameter space used in this search, and how the signal cross-section and kinematic distributions vary as we scan the parameter space; and Sect. 7.3 lists the Monte Carlo (MC) simulated samples for the signal model.

7.1 Introduction

As described in Chap. 6, there are a number of production mechanisms for DM plus Higgs, including either a contact operator in the EFT approach, or simplified models with a minimum number of renormalizable interactions. We consider here a simplified model with a Z' gauge boson and two-Higgs-doublets [1], where the DM is coupled to the heavy pseudoscalar Higgs boson.

7.1.1 Z'-2HDM Model

We consider a simplified model with renormalizable interactions where the relevant states may be produced on-shell. In light of the fact that high-dimension operators

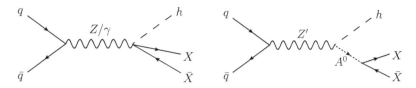

Fig. 7.1 Production mechanisms for dark matter plus Higgs through (*left*) a contact operator coupling dark matter to Zh or γh, or (*right*) a new Z' coupled to a two Higgs doublet model, where the new pseudoscalar A decays primarily to dark matter

are challenging to be UV-completed, it is more straightforward to generalize the process, as illustrated in Fig. 7.1. If the intermediate Z is instead a new Z' gauge boson, resonant production is possible. The Z' decays into a Higgs boson plus an intermediate state, which then decays into a pair of DM particles. For the "Higgs-portal" type of DM production, where the DM is directly coupled to a Higgs boson, at low DM mass below $m_h/2$, the constraints from the upper limit on Higgs invisible decay branching ratio are much stronger than the sensitivity of a direct search of this kind; at higher DM mass above $m_h/2$ where $h \to$ inv. constraints no longer apply, direct detection experiments are usually much more sensitive. Hence, we adopt a two-Higgs-doublet extension to the standard model (2HDM) [2], where $Z' \to hA$, and A is a heavy pseudoscalar with a large branching ratio to dark matter.

In this model, which we call Z'-2HDM, the gauge symmetry of the SM is extended by a $U(1)_{Z'}$, with a new massive Z' gauge boson. We assume that this sector also contains a SM singlet scalar ϕ that leads to spontaneous breaking of the symmetry and a Z' mass at a scale above electroweak symmetry breaking. For simplicity, we assume generation-independent charges for the fermions and that only the right-handed quarks u_R are charged.[1] This allows LHC production of the Z', but since the leptons are neutral, avoids potentially stringent constraints from searches for dilepton resonances.

For the Higgs sector, we use a type-2 two-Higgs-doublet model [2], where Φ_u couples to up-type quarks and Φ_d couples to down-type quarks and leptons:

$$-\mathcal{L} \supset y_u Q \tilde{\Phi}_u \bar{u} + y_d Q \Phi_d \bar{d} + y_e L \Phi_d \bar{e} + \text{h.c.} \tag{7.1}$$

After electroweak symmetry breaking, the Higgs doublets attain vevs v_u and v_d, and in unitary gauge the doublets are parametrized as

$$\Phi_d = \frac{1}{\sqrt{2}} \begin{pmatrix} -\sin\beta\, H^+ \\ v_d - \sin\alpha\, h + \cos\alpha\, H - i\sin\beta\, A \end{pmatrix},$$

---

[1] Anomaly cancellation can be achieved with a pair of colored triplet fields which are singlets with respect to $SU(2)_L$: $\psi_L(Q_z = 0, Y = -2/3)$ and $\psi_R(Q_z = -z_u, Y = -2/3)$ where z_u is the Z' charge of u_R.

7.1 Introduction

$$\Phi_u = \frac{1}{\sqrt{2}} \begin{pmatrix} \cos\beta\, H^+ \\ v_u + \cos\alpha\, h + \sin\alpha\, H + i\cos\beta\, A \end{pmatrix}, \qquad (7.2)$$

where h, H are neutral CP-even scalars and A is a neutral CP-odd scalar. Furthermore, $\tan\beta \equiv v_u/v_d$, and α is the mixing angle that diagonalizes the $h - H$ mass squared matrix.

Here h is assumed to correspond to the observed Higgs boson with $m_h \sim 125$ GeV. The remaining scalars $H, A, H^\pm$ are assumed to have masses around or above 300 GeV, in accordance with $b \to s\gamma$ constraints [2]. We further take $\alpha = \beta - \pi/2$, the alignment limit where h has SM-like couplings to fermions and gauge bosons [3], and $\tan\beta \geq 0.3$ based on the perturbativity requirement of the Higgs-top yukawa coupling [4].

7.1.2 *Z′ Constraints*

In this model, the Higgs vevs lead to $Z - Z'$ mass mixing. Diagonalizing the gauge boson mass matrix, the tree-level masses of the Z and Z' bosons are given by

$$\begin{aligned} m_Z^2 &\approx (m_Z^0)^2 - \epsilon^2 \left[(m_{Z'}^0)^2 - (m_Z^0)^2\right] \\ m_{Z'}^2 &\approx (m_{Z'}^0)^2 + \epsilon^2 \left[(m_{Z'}^0)^2 - (m_Z^0)^2\right], \end{aligned} \qquad (7.3)$$

where $(m_Z^0)^2 = g^2(v_d^2 + v_u^2)/(4\cos^2\theta_w)$ and $(m_{Z'}^0)^2 = g_z^2(z_d^2 v_d^2 + z_u^2 v_u^2 + z_\phi^2 v_\phi^2)$ are the mass-squared values in the absence of mixing. The result above is accurate to order ϵ^2, where ϵ is a small mixing parameter given by

$$\begin{aligned} \epsilon &\equiv \frac{1}{m_{Z'}^2 - m_Z^2} \frac{g g_z}{2\cos\theta_w}(z_d v_d^2 + z_u v_u^2) \\ &= \frac{(m_Z^0)^2}{m_{Z'}^2 - m_Z^2} \frac{2 g_z \cos\theta_w}{g} z_u \sin^2\beta. \end{aligned} \qquad (7.4)$$

The $Z - Z'$ mixing leads to a modification to the Z mass, as shown in Eq. (7.3), which affects the relation between the W and Z masses. This is expressed as a deviation of the ρ_0 parameter away from unity:

$$\rho_0 = 1 + \epsilon^2 \left(\frac{m_{Z'}^2 - m_Z^2}{m_Z^2} \right), \qquad (7.5)$$

Current precision electroweak global fits constrain $\rho_0 = 1.0004^{+0.0003}_{-0.0004}$ [5], which translates to an approximate 95% confidence level (CL) upper limit

$$\rho_0 \leq 1.0009. \qquad (7.6)$$

Combining the expressions in Eqs. (7.4)–(7.6), the upper limit on g_z from the ρ_0 constraint can be expressed as a function of $\tan \beta$ and $m_{Z'}$:

$$g_z \leq 0.03 \times \frac{g}{\cos \theta_w \sin^2 \beta} \times \sqrt{\frac{m_{Z'}^2 - m_Z^2}{m_Z^2}}. \tag{7.7}$$

This is shown in Fig. 7.2.

On the other hand, as Z' decays primarily to $q\bar{q}$, there are additional constraints from searches for dijet resonances at collider experiments. Results from Tevatron and LHC studies are used to derive 95 % CL upper limits on g_z, also shown in Fig. 7.2. The dijet constraints are calculated in a model-independent form in terms of fiducial cross-section (cross-section times acceptance) for a narrow resonance decaying to $q\bar{q}$, which is a valid approximation for this model assuming the Z' width is fixed for the most part by its decay to quarks and there isn't a significant width for Z' to decay to other new fermionic states.

As shown in Fig. 7.2, for Z' masses below ~ 1.3 TeV and larger $\tan \beta$, the ρ_0 constraint on g_z is stronger than dijet limits, while for $\tan \beta \lesssim 0.6$, the dijet constraints dominate even at low Z' masses. In signal yield calculation and sensitivity projections, we scale the signal production cross-section to the maximum g_z allowed at 95 % CL for the given $m_{Z'}$ and $\tan \beta$, as tabulated in Table 7.1.

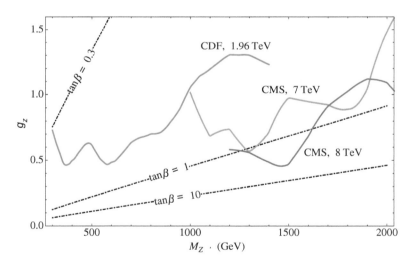

Fig. 7.2 95 % CL bounds on the Z' coupling g_z as a function of $m_{Z'}$. The *dashed lines* are upper bounds from ρ_0 parameter constraints on $Z - Z'$ mixing, for three values of $\tan \beta = 0.3, 1, 10$. Upper limits from dijet resonance searches at the Tevatron and at the LHC are also shown

7.1 Introduction

Table 7.1 Upper limit for g_z at 95 % CL for the given $m_{Z'}$ and $\tan\beta$ in the parameter space for this analysis

$\tan\beta$	0.3	0.5	1	3	5	10
$m_{Z'} = 600\,\text{GeV}$	0.45	0.45	0.25	0.16	0.15	0.15
$m_{Z'} = 800\,\text{GeV}$	0.7	0.7	0.35	0.22	0.21	0.2
$m_{Z'} = 1000\,\text{GeV}$	1.0	1.0	0.4	0.27	0.25	0.25
$m_{Z'} = 1200\,\text{GeV}$	0.55	0.55	0.5	0.33	0.31	0.3
$m_{Z'} = 1400\,\text{GeV}$	0.45	0.45	0.45	0.38	0.36	0.35

7.1.3 "Mono-Higgs" Signal

In the $h \to b\bar{b}$ decay channel, the signature of the signal we are searching for is missing transverse momentum (E_T^{miss}) recoiling against two b jets. In the Z'-2HDM model considered here, DM production is achieved via $Z' \to hA, h \to b\bar{b}, A \to \chi\chi$. The decay width of Z' for this to leading order in ϵ is

$$\Gamma_{Z' \to hA} = (g_z \cos\alpha \cos\beta)^2 \frac{|p|}{24\pi} \frac{|p|^2}{m_{Z'}^2}. \tag{7.8}$$

The center of mass momentum for the decay products is $|p| = \frac{1}{2m_{Z'}} \lambda^{1/2}(m_{Z'}^2, m_h^2, m_A^2)$, where λ is the Källen triangle function. This decay width is suppressed by $1/\tan^2\beta$ in the limit of large $\tan\beta$.

There is an additional source of $h + E_T^{\text{miss}}$ in the Z'-2HDM model, when Z' decays to a Higgs boson and a Z boson, and the Z boson decays invisibly. The decay width is

$$\Gamma_{Z' \to hZ} = (g_z \cos\alpha \sin\beta)^2 \frac{|p|}{24\pi} \left(\frac{|p|^2}{m_{Z'}^2} + 3\frac{m_Z^2}{m_{Z'}^2} \right). \tag{7.9}$$

For most of the parameter space we probe, the $Z' \to hA$ DM production mode is the dominant source of $h + E_T^{\text{miss}}$, but the contribution from $Z' \to hZ$ becomes more significant at small m_Z' due to the $m_Z^2/m_{Z'}^2$ term, and at large $\tan\beta$ when $Z' \to hA$ is suppressed. Both Z' decay modes are considered in this analysis, as the increased cross-section of the $h + E_T^{\text{miss}}$ signature allows us to probe a larger parameter space of this model. The two processes share very similar kinematics, as shown in Fig. 7.3: $Z' \to hZ$ shows a slightly more energetic spectrum due to the smaller Z mass compared to A mass.

7.1.4 Dark Matter Coupling to the Higgs Sector

In the Z'-2HDM model, we assume that the heavy pseudoscalar Higgs, A, has a large branching ratio to DM: when A mass is below twice of top quark mass, A decays almost exclusively to DM; when A mass is above that, there would be additional

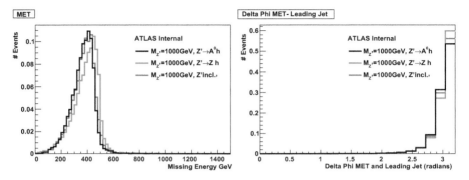

Fig. 7.3 Kinematic distributions of $Z' \to Ah$ exclusive production, $Z' \to Zh$ exclusive production and Z' inclusive production for $m_{Z'} = 1000\,\text{GeV}$, $m_A = 300\,\text{GeV}$, and $\tan\beta = 1$

decay modes of $A \to t\bar{t}$ which reduces the signal cross-section. There are a number of theoretical possibilities that allow the desired large coupling between dark matter to the Higgs sector in this model, the details of which vary depending on the type of DM, taking into account direct detection and relic density considerations.

For fermionic DM, one possibility is a pseudoscalar interaction with a singlet–doublet DM coupling to the down-type Higgs. In this model, a singlet S and electroweak doublets $D_{1,2}$ (all singlets under $U(1)_{Z'}$) are introduced, with a Lagrangian

$$-\mathcal{L} \supset \frac{1}{2}M_S^2 S^2 + M_D D_1 D_2 + y_1 S D_1 \Phi_d + y_2 S \Phi_d^\dagger D_2 + \text{h.c.}$$

By requiring DM mass to be at least half of Higgs boson mass, constraints from the invisible decay width of the Higgs boson can be avoided. Though elastic scattering off quarks can proceed via the exchange of h or H leading to restrictive direct detection constraints, in parts of the parameter space near the "blind spot" where the coupling through the Higgs is suppressed ($\tan\theta < 0$, where $y_1 = y\cos\theta$, $y_2 = y\sin\theta$), the direct detection cross-sections are small. It is hence possible to obtain large branching ratios of A^0 to DM while satisfying LUX constraints [6].

For scalar DM, we can consider a complex scalar field X, written as $X = \frac{1}{\sqrt{2}}(X_1 + iX_2)$, where X_1 and X_2 are two real fields, and X is a SM singlet and has $U(1)_{Z'}$ charge $-1/4$. The renormalizable interactions of the DM with the Higgs sector are

$$\mathcal{L} \supset \left(\lambda_{dd}|\Phi_d|^2 + \lambda_{uu}|\Phi_u|^2\right)|X|^2 + \left(\lambda_{du}\Phi_d^\dagger\Phi_u X^2 + \text{h.c.}\right), \tag{7.10}$$

with all couplings taken to be real. In this case, the direct detection limits from LUX can be satisfied with the couplings λ_{dd}, λ_{uu}, and λ_{du} at the 10 % level [7] and a DM mass scale of $\sim 100\,\text{GeV}$.

7.2 Parameter Space and Kinematic Dependencies

From Sect. 7.1, we know there are a total of five "free" parameters in this model, $\tan \beta (\equiv v_u/v_d)$, the Z' gauge coupling g_z, $m_{Z'}$, m_A, and dark matter mass m_{DM}. To study the signal production and kinematic dependencies on these parameters, we produced signal samples varying each of the five parameters generated at parton level in MadGraph [8], using PYTHIA8 [9] for the parton showering and hadronization, and detector simulation with DELPHES [10] or official ATLAS ATLFAST-II simulation [11]. The results using DELPHES simulation are consistent with signals using official ATLAS simulation ATLFAST-II [11] as listed in Sect. 7.3.

As g_z is the coupling strength of the Z' gauge boson, variations in its value do not lead to any kinematic changes. The signal cross-section scales as $(g_z)^2$, which can be inferred from Eqs. (7.8) and (7.9).

In the DM production mode of $Z' \to hA$, $A \to \chi\chi$, as the DM pair are produced from resonant decay of A, there is minimal kinematic changes by varying m_{DM} as long as $m_{\mathrm{DM}} < m_A/2$ so that A production is on-shell, as shown in Fig. 7.4.

As seen in Fig. 7.5, variations of $\tan \beta$ do not lead to changes in kinematic distributions. The signal cross-section scales as $1/\tan^2 \beta$ for $Z' \to Ah$ [Eq. (7.8)], and $\sin^4 \beta$ for $Z' \to Zh$ [Eq. (7.9)]. The signal samples used in the analysis are produced with a fixed $\tan \beta$ value of 1 and the plots shown in Fig. 7.5 for the $Z' \to Ah$ process.

Summing up the three points above, signal events have been produced for a fixed $g_z = 0.8$, $\tan \beta = 1$, and $m_{\mathrm{DM}} = 100 \, \mathrm{GeV}$. For these values, we scan the 2-D parameter space of $m_{Z'}$, m_A with $m_{Z'} = 600, 800, 1000, 1200, 1400 \, \mathrm{GeV}$, and $m_A = 300, 400, 500, 600, 700, 800 \, \mathrm{GeV}$ when fulfilling $m_A < m_{Z'} - m_h$, for a total of 24 grid points. Kinematic shifts as $m_{Z'}$ is varied at fixed m_A are shown in Fig. 7.6, and the dependency on m_A is shown in Fig. 7.7. Both behave as expected: when m'_Z is much larger than m_A, the Higgs boson is more energetic, and the two b-quark jets from the Higgs boson decay have a larger p_T and smaller angular separation; the E_T^{miss} recoils against the Higgs boson decay products, and has a harder spectrum as well.

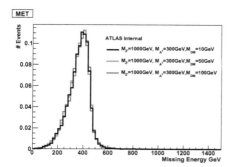 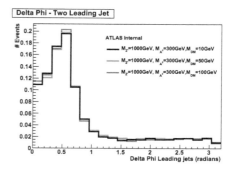

Fig. 7.4 Kinematic distributions of the $Z' \to Ah$ signal process varying m_{DM}: minimal kinematic dependency on m_{DM} as expected when A is produced on-shell. Plots shown for $m_{Z'} = 1000 \, \mathrm{GeV}$, $m_A = 300 \, \mathrm{GeV}$

7 Dark Matter + Higgs($\to b\bar{b}$): Z'-2HDM Simplified Model

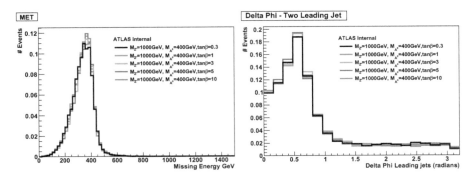

Fig. 7.5 Kinematic distributions of the $Z' \to Ah$ signal process varying $\tan\beta$: no kinematic dependency on $\tan\beta$ as expected

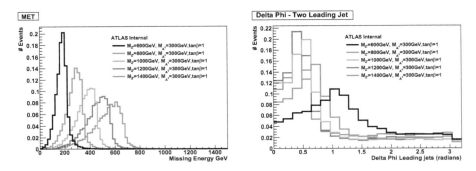

Fig. 7.6 Kinematic distributions of the $Z' \to Ah$ signal process varying $m_{Z'}$: behavior as expected. Plots shown for $m_{DM} = 100\,\text{GeV}$ and $m_A = 300\,\text{GeV}$

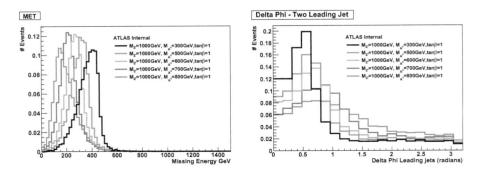

Fig. 7.7 Kinematic distributions of the $Z' \to Ah$ signal process varying m_A: behavior as expected. Plots shown for $m_{Z'} = 1000\,\text{GeV}$ and $m_{DM} = 100\,\text{GeV}$

7.3 Simulated Signal Samples

Monte Carlo (MC) simulated event samples are used to model the Z'-2HDM signals. The simulated samples are processed with a fast simulation of the response of the electromagnetic and hadronic calorimeters of the ATLAS detector [11]. The results based on fast simulations are validated against samples processed with a full ATLAS detector simulation [12] and the difference is found to be negligible. The simulated samples are further processed with a simulation of the trigger system. Both the simulated events and the data are reconstructed and analyzed with the same analysis chain, using the same event selection criteria.

Signal samples are generated with MADGRAPH [8], interfaced to PYTHIA8 using the AU2 parameter settings (tune) [13] for parton showering, hadronization, and underlying event simulation. The Higgs boson mass is fixed to 125 GeV. The parton distribution functions (PDF) is defined with the central value in the MSTW2008LO leading-order (LO) PDF set [14]. The decay process of $Z' \to hA$, $h \to bb$, $A \to \chi\chi$ is done in Madgraph: as the Higgs decay ratio is calculated at loop level, which could not be performed by Madgraph in an LO calculation, per theory recommendation, the $h \to bb$ is set to an automated value of ~ 0.92 in Madgraph, and the signal cross-section is scaled to the loop-level calculation of the $h \to bb$ branching ratio of ~ 0.57.

Samples are produced with Z' mass between 600 and 1400 GeV, A mass between 300 and 800 GeV (where kinematically allowed), and DM mass between 10 and 200 GeV but always less than half the A mass. In addition, $Z' \to Zh$ samples are produced for Z' mass between 600 and 1400 GeV. All signals are produced with $g_z = 0.8$ and $\tan\beta = 1$: in signal yield calculation and sensitivity projections, we scale the signal production cross-section to the maximum g_z allowed at 95 % CL for the given $m_{Z'}$ and $\tan\beta$ (Fig. 7.2, Table 7.1), as well as the cross-section dependence on $\tan\beta$ as given in Eqs. (7.8) and (7.9).

The signal samples are listed in Tables 7.2, 7.3, and 7.4. The K-factors are typically the ratio of the NLO to LO cross-section for a given process. The signals are computed at tree-level; with the heavy Z' and small g_z coupling, the loop corrections are small compared with other signal systematics, so we use 1.0 for K-factors here. The filter efficiency ε is applied at generated parton level in simulation; in this case there are no specific filters applied to the generated signal samples, so the value is 1.

References

1. A. Berlin, T. Lin, L.-T. Wang, Mono-Higgs detection of dark matter at the LHC. J. High Energy Phys. **6**, 078 (2014)
2. G.C. Branco, P.M. Ferreira, L. Lavoura, M.N. Rebelo, M. Sher, J.P. Silva, Theory and phenomenology of two-Higgs-doublet models. Phys. Rep. **516**, 1–102 (2012)
3. N. Craig, J. Galloway, S. Thomas, Searching for signs of the second Higgs doublet (2013)

Table 7.2 $Z' \to Ah$ signal samples used for the analysis, varying $m_{Z'}$ and m_A with $g_z = 0.8$, $\tan\beta = 1$, and $m_{DM} = 100\,\text{GeV}$

ID	$m_{Z'}$ (GeV)	m_A (GeV)	Generator	σ (pb)	K-factor	ε (%)	$N_{\text{gen}}^{\text{weighted}}$
203850	600	300	MadGraph+Pythia8	1.55E−01	1.0	100	20,000
203851	600	400	MadGraph+Pythia8	2.18E−02	1.0	100	20,000
203852	800	300	MadGraph+Pythia8	8.30E−02	1.0	100	20,000
203853	800	400	MadGraph+Pythia8	2.72E−02	1.0	100	20,000
203854	800	500	MadGraph+Pythia8	1.09E−02	1.0	100	20,000
203855	800	600	MadGraph+Pythia8	2.98E−03	1.0	100	20,000
203856	1000	300	MadGraph+Pythia8	3.74E−02	1.0	100	20,000
203857	1000	400	MadGraph+Pythia8	1.53E−02	1.0	100	20,000
203858	1000	500	MadGraph+Pythia8	8.91E−03	1.0	100	20,000
203859	1000	600	MadGraph+Pythia8	4.89E−03	1.0	100	20,000
203860	1000	700	MadGraph+Pythia8	2.21E−03	1.0	100	20,000
203861	1000	800	MadGraph+Pythia8	7.05E−04	1.0	100	20,000
203862	1200	300	MadGraph+Pythia8	1.70E−02	1.0	100	20,000
203863	1200	400	MadGraph+Pythia8	7.65E−03	1.0	100	20,000
203864	1200	500	MadGraph+Pythia8	5.14E−03	1.0	100	20,000
203865	1200	600	MadGraph+Pythia8	3.52E−03	1.0	100	20,000
203866	1200	700	MadGraph+Pythia8	2.25E−03	1.0	100	20,000
203867	1200	800	MadGraph+Pythia8	1.27E−03	1.0	100	20,000
203868	1400	300	MadGraph+Pythia8	8.00E−03	1.0	100	20,000
203869	1400	400	MadGraph+Pythia8	3.79E−03	1.0	100	20,000
203870	1400	500	MadGraph+Pythia8	2.75E−03	1.0	100	20,000
203871	1400	600	MadGraph+Pythia8	2.09E−03	1.0	100	20,000
203872	1400	700	MadGraph+Pythia8	1.58E−03	1.0	100	20,000
203873	1400	800	MadGraph+Pythia8	1.06E−03	1.0	100	20,000

The columns from left to right describe MC generation ID, $m_{Z'}$, m_A, the MC generator used, the sample cross-section in pb, the NLO/LO K-factor (multiplicative to the cross-section), the filter efficiency of the requested sample within the ATLAS simulation, and the effective number of events for normalization

Table 7.3 $Z' \to Ah$ signal samples used for the analysis, varying m_{DM} with $g_z = 0.8$, $\tan\beta = 1$

ID	$m_{Z'}$ (GeV)	m_A (GeV)	m_{DM} (GeV)	Generator	σ (pb)	K-factor	ε (%)	$N_{\text{gen}}^{\text{weighted}}$
203874	1000	300	10	MadGraph+Pythia8	3.76E−02	1.0	100	20,000
203875	1000	300	50	MadGraph+Pythia8	3.75E−02	1.0	100	20,000
203876	1200	600	10	MadGraph+Pythia8	3.64E−03	1.0	100	20,000
203877	1200	600	20	MadGraph+Pythia8	3.07E−03	1.0	100	20,000

The columns from left to right describe MC generation ID, $m_{Z'}$, m_A, m_{DM}, the MC generator used, the sample cross-section in pb, the NLO/LO K-factor (multiplicative to the cross-section), the filter efficiency of the requested sample within the ATLAS simulation, and the effective number of events for normalization

Table 7.4 $Z' \to Zh$ exclusive samples produced for the analysis, varying $m_{Z'}$ with $g_z = 0.8$, $\tan\beta = 1$

ID	$m_{Z'}$ (GeV)	Generator	σ (pb)	K-factor	ε (%)	$N_{\text{gen}}^{\text{weighted}}$
203878	600	MadGraph+Pythia8	1.15E−01	1.0	100	20,000
203879	800	MadGraph+Pythia8	3.21E−02	1.0	100	20,000
203880	1000	MadGraph+Pythia8	1.13E−02	1.0	100	20,000
203881	1200	MadGraph+Pythia8	4.54E−03	1.0	100	20,000
203882	1400	MadGraph+Pythia8	2.00E−03	1.0	100	20,000

The columns from left to right describe MC generation ID, $m_{Z'}$, the MC generator used, the sample cross-section in pb, the NLO/LO K-factor (multiplicative to the cross-section), the filter efficiency of the requested sample within the ATLAS simulation, and the effective number of events for normalization

4. A. Azatov, S. Chang, N. Craig, J. Galloway, Higgs fits preference for suppressed down-type couplings: implications for supersymmetry. Phys. Rev. **D86**, 075033 (2012)
5. J. Beringer et al., Review of particle physics (RPP). Phys. Rev. **D86**, 010001 (2012)
6. D.S. Akerib et al., First results from the LUX dark matter experiment at the Sanford Underground Research Facility. Phys. Rev. Lett. **112**, 091303 (2014)
7. X.-G. He, J. Tandean, Low-mass dark-matter hint from CDMS II, Higgs boson at the LHC, and Darkon models. Phys. Rev. **D88**, 013020 (2013)
8. J. Alwall, M. Herquet, F. Maltoni, O. Mattelaer, T. Stelzer, MadGraph 5: going beyond. J. High Energy Phys. **06**, 128 (2011)
9. T. Sjöstrand, S. Mrenna, P.Z. Skands, A brief introduction to PYTHIA 8.1. Comput. Phys. Commun. **178**, 852–867 (2008)
10. J. de Favereau, C. Delaere, P. Demin, A. Giammanco, V. Lemaître et al., DELPHES 3, A modular framework for fast simulation of a generic collider experiment (2013)
11. ATLAS Collaboration, The simulation principle and performance of the ATLAS fast calorimeter simulation FastCaloSim. (ATL-PHYS-PUB-2010-013) (2010)
12. ATLAS Collaboration, The ATLAS experiment at the CERN large hadron collider. J. Instrum. **3**, S08003 (2008)
13. ATLAS Collaboration, Summary of ATLAS Pythia 8 tunes. (ATL-PHYS-PUB-2012-003) (2012)
14. A.D. Martin, W.J. Stirling, R.S. Thorne, G. Watt, Parton distributions for the LHC. Eur. Phys. J. **C63**, 189–285 (2009)

Chapter 8
Dark Matter + Higgs($\to b\bar{b}$): Physics Objects

This chapter describes the physics objects used in this analysis, including the lists of samples from data and simulated background processes, and the physical and kinematic variables constructed that are used to identify the events and separate background from signal. The signal model and simulated signal events are presented in Chap. 7. The event selections based on values of these variables are discussed in Chap. 9.

This chapter is organized as follows: Sect. 8.1.1 presents the data samples used in this analysis, while the simulated samples of the background processes are given in Sect. 8.1.2; Sect. 8.2 describes in detail the E_T^{miss} trigger, which is the primary trigger used in this analysis, and also lists the other lepton and jet triggers used; and Sect. 8.3 describes the physical and kinematic variables relevant to this analysis, their definition, and the selection of these objects.

8.1 Data and Simulated Background Processes

Events from both pp collision data recorded by the ATLAS detector and Monte Carlo (MC) simulation are used in this analysis. All samples used are in the format of SUSY D3PDs with tag p1328 originating from official ATLAS AOD productions.

8.1.1 Data Sample

The search in this thesis is based on pp collision data collected by the ATLAS detector during LHC Run I in 2012 with center of mass energy $\sqrt{s} = 8$ TeV, corresponding to a total integrated luminosity of 20.3 fb^{-1}. These datasets are

Table 8.1 Summary of the 2012 ATLAS datasets used for this analysis

Year	$\sqrt{s}$	Peak luminosity (cm^{-1} s^{-1})	Pile-up $\langle\mu\rangle$	Data quality efficiency	Integrated luminosity (fb^{-1})
2012	8 TeV	7.73×10^{33}	20.3	95 %	20.3

Table 8.2 Details of the 2012 ATLAS datasets used for this analysis, listed by their data taking periods

Period	Dates	Run numbers	L (pb^{-1})
A	Apr-04 : Apr-20	200804 : 216432	794.80
B	May-01 : Jun-18	202660 : 205113	5104.55
C	Jul-01 : Jul-24	206248 : 207397	1408.33
D	Jul-24 : Aug-23	207447 : 209025	3296.18
E	Aug-23 : Sep-17	209074 : 210308	2531.50
G	Sep-26 : Oct-08	211522 : 212272	1276.43
H	Oct-13 : Oct-26	212619 : 213359	1447.39
I	Oct-26 : Nov-02	213431 : 213819	1019.28
J	Nov-02 : Nov-26	213900 : 215091	2603.29
L	Nov-30 : Dec-06	215414 : 215643	824.12
A–L	Apr-04 : Dec-06	200804 : 215643	20,323.9

summarized in Table 8.1. Events are only used when all relevant parts of the detector are working nominally.

The datasets were recorded by the ATLAS detector between April and December of 2012 and correspond to the data taking periods A, B, C, D, E, G, H, I, J, and L, which comprise the run numbers 200804–215643. Table 8.2 gives an overview of the data taking periods, their corresponding run numbers, and integrated luminosities.

For the signal region and 0-lepton validation regions, data from the JetTauEtmiss data stream selected with an unprescaled E_T^{miss} trigger is used; details of the E_T^{miss} trigger are given in Sect. 8.2.1. The electron control region and γ + jets data are selected with either unprescaled single electron triggers or a γ trigger from the EGamma data stream, giving the same luminosity as the JetTauEtmiss stream. The muon control region data is from the muon data stream, selected with the dedicated muon triggers, yielding the same luminosity.

8.1.2 Simulated Background Samples

Monte Carlo (MC) simulated event samples are used to model different background processes. Effects of multiple proton–proton interactions (pile-up) as a function of the instantaneous luminosity are taken into account by overlaying simulated minimum-bias events generated with PYTHIA8 [1] onto the hard-scattering process, such that the distribution of the average number of interactions per bunch crossing in the Monte Carlo simulated samples matches that in the data. The simulated

8.1 Data and Simulated Background Processes

samples are processed either with a full ATLAS detector simulation [2] based on the GEANT4 program [3], or a fast simulation of the response of the electromagnetic and hadronic calorimeters [4]. The results based on fast simulations are validated against fully simulated samples and the difference is found to be negligible. The simulated samples are further processed with a simulation of the trigger system. Both the simulated events and the data are reconstructed and analyzed with the same analysis chain, using the same event selection criteria.

The dominant $Z(\rightarrow \nu\bar{\nu})$+jets background is determined from data (Sect. 10.3.2), and samples simulated with SHERPA [5] for $Z(\rightarrow \nu\bar{\nu})$+jets, $Z(\rightarrow \ell\ell)$+jets, and γ+jets are also used in the calculation process. The $W(\rightarrow \ell\nu)$+jets processes are generated with SHERPA and are normalized using data as described in Sect. 10.2.1. All the SHERPA samples are generated using the CT10 PDF set [6]. The SHERPA samples are produced in slices of Z boson p_T as well to ensure sufficient statistics up to high E_T^{miss}. Moreover, the samples were produced in three exclusive heavy-flavor compositions: veto on bottom and charm quarks, allow for charm quarks but veto bottom quarks, allow for bottom quarks only.

The $t\bar{t}$ background is generated with POWHEG-BOX [7] interfaced with PYTHIA6 and the PERUGIA 2011C tune [8]. Single top quark production in the s- and Wt-channels are produced with MC@NLO [9–11] interfaced with JIMMY [12], while the t-channel is produced with ACERMC [13] interfaced with PYTHIA6. The Diagram Removal scheme [14] is used in the single top quark production in the Wt-channel to remove potential interference with $t\bar{t}$ production. A top quark mass of 172.5 GeV is used consistently. The cross-sections of the $t\bar{t}$ and single-top-quark processes are determined at next-to-next-to-leading order (NNLO) in QCD including resummation of next-to-next-to-leading logarithmic (NNLL) soft gluon terms with Top++2.0 [15–21]. The normalization and uncertainties are calculated using the PDF4LHC prescription [22] with the MSTW2008 68 % CL NNLO [23, 24], CT10 NNLO [6, 25], and NNPDF2.3 [26] PDF sets. Additional kinematic-dependent corrections to the $t\bar{t}$ sample and normalizations determined from data are described in Sect. 10.2.2.

Diboson (ZZ, WW, and WZ) production is simulated with HERWIG [27] interfaced to JIMMY. The diboson samples are normalized to calculations at next-to-leading order (NLO) in QCD performed using MCFM [28].

The multijet background is estimated from data (Sect. 10.3.1), with samples simulated with PYTHIA8 used for validation in the control regions.

Standard model Higgs events when Higgs is produced in association with a vector boson and decays to a pair of b-quarks, i.e., $VH(b\bar{b})$ events, are considered as an additional background in the signal region (it is negligible in the CRs). For SM production of Zh and Wh, PYTHIA8 is used with CTEQ6L1 PDFs, and the samples were normalized to total cross-sections calculated at NLO [29], and NNLO [30] in QCD, respectively, with NLO electroweak corrections [31] in both cases.

Table 8.3 summarizes the various event generators and parton distribution function (PDF) sets, as well as parton shower and hadronization software used for the analyses presented in this thesis.

Table 8.3 Summary of MC event generators, PDF sets, and parton shower and hadronization models utilized in the analyses for both the signal and background processes

Model/Process	Generator	PDF	Shower model
Z'-2HDM	MADGRAPH v1.5.1	MSTW2008LO	PYTHIA v8.175
EFT models	MADGRAPH v1.5.1	CTEQ6L1	PYTHIA v8.175
$W/Z/\gamma$+jets	SHERPA v1.4.3	CT10	SHERPA v1.4.3
$t\bar{t}$	POWHEG-BOX v1.0 r2129	CT10	PYTHIA v6.427
Single top (s-ch., Wt)	MC@NLO v3.31	CT10	JIMMY v4.31
Single top (t-ch.)	ACERMC v3.8	CTEQ6L1	PYTHIA v6.426
$WW/WZ/ZZ$	HERWIG v6.520	CTEQ6L1	JIMMY v4.31
$q\bar{q} \to Vh$	PYTHIA v8.175	CTEQ6L1	PYTHIA v8.175
$gg \to Zh$	POWHEG r2330.3	CT10	PYTHIA v8.175
Multijet	PYTHIA v8.160	CT10	PYTHIA v8.160

8.2 Trigger

The primary data sample is selected using an E_T^{miss} trigger yielding hadronic final states with missing energy, as discussed in detail in Sect. 8.2.1. Lepton and photon triggers are also used to select events used in the control regions and validation regions to study and test the modeling of background processes, as described in Sect. 8.2.2.

8.2.1 E_T^{miss} Trigger

An E_T^{miss} trigger is used in this analysis to select events with fully hadronic final states. The threshold is 60 GeV at the L1 trigger and 80 GeV at the HLT. The trigger efficiency is above 98 % for events passing the full offline selection across the full E_T^{miss} range considered in this analysis.

8.2.1.1 Definition and Implementation

The E_T^{miss} trigger used in this analysis is denoted as `EF_xe80_tclcw`. Details about the implementation of the E_T^{miss} trigger can be found in [32]. This trigger chain has the following thresholds:

- L1 $E_T^{miss} > 60$ GeV
- L2 $E_T^{miss} > 65$ GeV
- EF $E_T^{miss} > 80$ GeV

At Level 1, E_T^{miss} is calculated from calorimeter information alone; Level 2 uses output from Level 1 plus information from the muon spectrometer and inner detectors; the EventFilter (EF) then applies a calibration scheme "Local Cluster Weighting" (LCW) [33]. The LCW calibration method classifies the topological calorimeter clusters as either electromagnetic or hadronic, using the energy density and the shape of the cluster; this reconstruction method improves the E_T^{miss} resolution of the E_T^{miss} triggers. The EF_xe80_tclcw trigger was used for all data taking periods of the 2012 run, and is not prescaled.

8.2.1.2 Trigger Efficiency in Data and Simulation

We study the efficiency of the EF_xe80_tclcw trigger as a function of E_T^{miss}, and compare the result between data and MC. The data sample used for the trigger study is the full 21 fb^{-1} 2012 muon stream data, which has both leptons (muons) and E_T^{miss} in the final state. A number of different MC samples simulated with the ALPGEN generator are used for comparison, as listed in Table 8.5. All the MC samples are scaled by cross-section corresponding to luminosity. Both data and simulated events are required to pass the selection criteria in Table 8.4 to select $W \to \mu\nu$ events.

Trigger efficiency is reconstructed as a function of the offline reconstructed E_T^{miss}, and the resulting efficiency "turn-on" curves are fit with a Fermi function [Eq. (8.1)].

$$f(x) = \frac{1}{1 + e^{(x-a)/b}} \qquad (8.1)$$

Table 8.4 Trigger selections to enrich $W \to \mu\nu$

Jets	
At least one jet, medium quality	
p_T(jet)	> 30 GeV
$|\eta$(jet)$|$	< 4.5
Leading jet p_T	> 100 GeV
Muons	
Trigger	EF_mu24i_tight
	EF_mu36_tight
Quality	Good muon, TightIso
	$E_{T\,cone}^{30}/p_T < 0.12$
$p_T(\mu)$	> 25 GeV
$|\eta(\mu)|$	< 2.5

The muons are required to be isolated using the calorimeter-based isolation requirement: a muon is considered well isolated when the sum of the energy deposits in topological clusters in the calorimeter in a cone of $\delta R = 0.3$ around the muon, but excluding the muon itself, i.e., $E_{T\,cone}^{30}$, is below 0.12 times the muon p_T

Table 8.5 Event yield and E_T^{miss} trigger efficiency values for data and simulated background samples using the EF_xe80_tclcw trigger

	Data	$W\mu\nu$	W + jets incl	Diboson	$t\bar{t}$	Simulation comb
Sample size	2.8×10^6	1.38×10^6	4.92×10^6	8.79×10^4	4.6×10^5	5.47×10^6
Before trig	2.41×10^6	1.34×10^6	1.73×10^6	3.12×10^4	3.12×10^5	2.07×10^6
After trig	2.01×10^6	1.28×10^6	1.66×10^6	3.00×10^4	3.72×10^5	1.96×10^6
Max. Eff. (%)	99.7	99.8	99.9	99.2	99.2	99.6
E_T^{miss} at 98 % (GeV)	151	131	133	136	200	145
χ^2/dof	0.34	0.09	0.11	0.40	0.35	0.12

8.2 Trigger

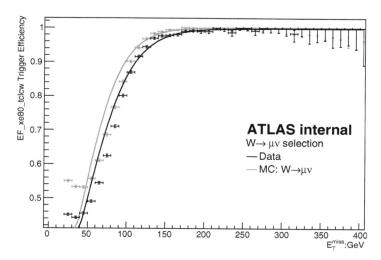

Fig. 8.1 EF_xe80_tclcw trigger turn-on curves as a function of E_T^{miss} for data (*black*) and $W(\to \mu\nu)$+jets exclusive process in simulation (*red*)

An error function [Eq. (8.2)] has also been used for cross-check, yielding compatible results with a slightly larger χ^2. Hence the Fermi function in Eq. (8.1) is used to interpret the results.

$$f(x) = \frac{2}{\sqrt{\pi}} \int_0^x e^{-t^2} \, dt \qquad (8.2)$$

A ~2 % discrepancy in efficiency curves between data and simulated $W(\to \mu\nu)$+jets exclusive sample is observed, as shown in Fig. 8.1. This difference is likely to come from events in data that passed the selection and are mis-reconstructed as $W \to \mu\nu$ events, instead of being the result of difference in trigger turn-on between data and simulation. This is confirmed by studying different simulated background processes individually and combined, as shown in Fig. 8.2, where the combined MC sample of W+jets inclusive, diboson, and $t\bar{t}$ events yields near identical trigger efficiency as data. The E_T^{miss} value where the turn-on curve reaches its plateau, the E_T^{miss} value at 98 % of highest efficiency, and the χ^2/dof are also calculated and tabulated in Table 8.5. The E_T^{miss} trigger reaches 98 % efficiency at 151 GeV in the data, and at 145 GeV in the combined background processes in simulation. The analysis adopts a baseline E_T^{miss} threshold 100 GeV, at which point and beyond data and combined MC samples have near perfect agreement.

The difference between the trigger efficiency curve from W+jets inclusive MC sample, and that of the data, as shown in Fig. 8.2a, is applied as a correction to all simulated events with E_T^{miss} below 180 GeV. In this analysis, most of the signal samples have a selection cut of E_T^{miss} above 200 GeV, where the triggers are fully efficient for all samples in discussion. For the two signal samples with

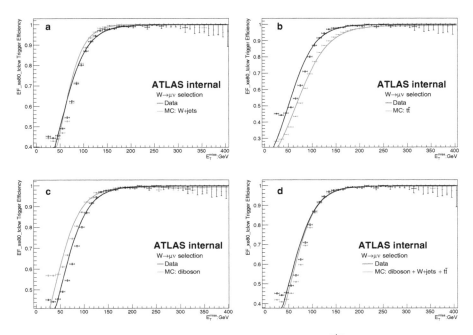

Fig. 8.2 EF_xe80_tclcw trigger turn-on curves as a function of E_T^{miss} for (**a**) W+jets inclusive process, (**b**) $t\bar{t}$, (**c**) Diboson, and (**d**) combined background processes in simulation

$m_{Z'} = 600$ GeV that use a selection cut of E_T^{miss} above 150 GeV, there is a small discrepancy in trigger turn-on in $t\bar{t}$ events compared with that in $W +$ jets events (the red curves in Fig. 8.2a, b) leading to an uncertainty of 6 % before the trigger reaches full efficiency for $t\bar{t}$ events, but as it only applies to a narrow region of E_T^{miss}, and the discrepancy is much smaller compared with other sources of systematic uncertainty in $t\bar{t}$ (Sect. 11.4), no additional systematic uncertainty is added for this case.

8.2.1.3 Kinematic and Pile-Up Dependencies

We further study properties of the EF_xe80_tclcw trigger in terms of possible correlations between the trigger efficiency and kinematic or pile-up conditions. The studies are done by comparing the trigger efficiency as a function of E_T^{miss} in events passing different requirements on various kinematic variables or relevant properties, and are performed for both data and the different background processes in simulation.

Trigger Efficiency and Leading Jet Transverse Momentum The dependence of the E_T^{miss} trigger efficiency on leading jet transverse momentum is studied by deducing the turn-on curves for various requirements on leading jet transverse

8.2 Trigger

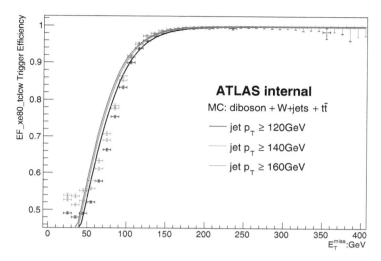

Fig. 8.3 Trigger turn-on curves for different jet momentum using the `EF_xe80_tclcw` trigger

momentum p_T(leading jet). The selections used are p_T(leading jet) > 120, 140, and 160 GeV. The trigger efficiency shows a weak dependency on the leading jet p_T in both data and MC, consistent with previous studies found in [34]. The trigger turn-on curves are shown in Fig. 8.3.

Trigger Efficiency and Number of Vertices The dependence of the trigger efficiency on the pile-up condition is also studied. Events are selected according to the number of reconstructed primary vertices[1] and divided into three subgroups: $N_{vtx} \leq 5$, $5 < N_{vtx} \leq 10$, and $N_{vtx} > 10$. Trigger efficiency curves are calculated for each subgroup and shown in Fig. 8.4. One observes a small dependency on the number of vertices, consistent with previous studies [34].

Trigger Efficiency and Jet Multiplicity We study the dependence of trigger efficiency on number of jets in the final state. Events are selected and divided into three subgroups according to the jet multiplicity: 1 jet, 2 jets, 3 and more jets. Trigger efficiency curves are obtained for each subgroup and as we can see in Fig. 8.5, the turn-on curves reach 98 % or full efficiency at slightly larger E_T^{miss} for events with larger jet multiplicities. This is consistent with previous analysis [34] and expected due to pile-up and possible mis-measurement at higher jet multiplicity. This could also partially explain the slightly lower turn-on curve for $t\bar{t}$ events which would have a larger jet multiplicity, as observed in Fig. 8.2. This analysis requires

---

[1] Proton–proton collision vertices are reconstructed requiring that at least five tracks with $p_T > 0.5$ GeV are associated with a given vertex.

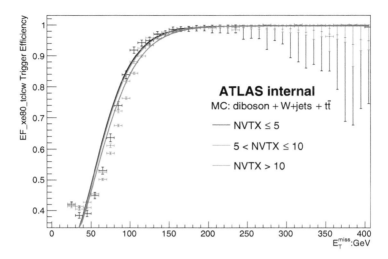

Fig. 8.4 EF_xe80_tclcw trigger efficiency dependency on N_{vtx}

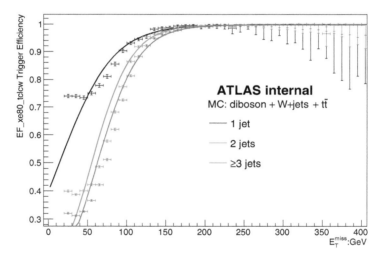

Fig. 8.5 EF_xe80_tclcw trigger efficiency dependency on jet multiplicity

2 or 3 jets in its final state, for which the change in trigger efficiency curve is small. This will also reduce the aforementioned small discrepancy in trigger turn-on between $t\bar{t}$ events and other sources of background.

8.2.2 Other Triggers

Muon triggers with transverse momentum thresholds at the HLT of 24 GeV for muons with surrounding inner-detector tracking activity below a predefined level, i.e., isolated muons [35], and 36 GeV for muons with no isolation requirement, are used to select muon data used for the estimation and validation of backgrounds in control regions. The efficiency of this trigger combination reaches its plateau at just under 25 GeV of muon p_T, after which the trigger efficiency is about 86 % (70 %) for muons traveling through the end-cap (barrel) regions of the detector.

A photon trigger with a transverse momentum threshold of 120 GeV at the HLT is used to select events with a high p_T prompt photon for data-driven $Z(\to \nu\bar{\nu})$+jets background estimation (Sect. 10.3.2). This trigger has almost full efficiency for photons with p_T greater than 125 GeV.

A set of jet triggers with different p_T thresholds are used in the data-driven method to estimate multijet background (Sect. 10.3.1).

8.3 Final-State Observables: Definition and Selection

The analysis uses a wide variety of physics objects that are reconstructed offline from properties recorded by the ATLAS detector, including jets which may be identified for the flavor of quarks it contains, electrons, muons, photons, and E_T^{miss}. Object reconstruction efficiencies in simulated events are corrected to reproduce the performance measured in the data, and their systematic uncertainties are detailed in Chap. 11. The definitions and selections of these objects are detailed in the following paragraphs.

Jets Jets are reconstructed at the electromagnetic (EM) scale using the anti-k_t algorithm with an angular coverage of $\Delta R = 0.4$. The input objects to the jet algorithm are three-dimensional topological clusters (topological calorimeter clusters) [36] built from energy deposits in calorimeter cells. Each topocluster is constructed from a seed calorimeter cell with $|E_{cell}| > 4\sigma$, where σ is the RMS of the noise of the cell. Neighboring cells are iteratively added to the topocluster if they have $|E_{cell}| > 2\sigma$. Finally, an outer layer of surrounding cells is added. In the jet reconstruction, each such calorimeter cluster is considered as a massless particle with energy $E = \sum E_{cell}$, with position at the energy-weighted barycenter of the cells in the cluster and originating from the geometrical center of the ATLAS detector. The four momentum of the uncalibrated, EM-scale jet is defined as sum of four momenta of the calorimeter clusters. The jet is then calibrated in three subsequent steps as outlined below. The calibration procedure is detailed in [37].

- Additional energy due to multiple proton–proton interactions within the same bunch crossing (pile-up) is subtracted using a correction measured in data [38];

- The position of the jet is corrected such that the jet direction points to the primary vertex of the interaction instead of the geometrical center of ATLAS;
- Finally, the energy and the position of the jet are corrected for instrumental effects (calorimeter non-compensation, additional dead material, and out-of-cone effects) and the jet energy scale is restored on average to that of the particles entering the calorimeter.

A jet is categorized as either "central" or "forward" depending on its η value. Jets selected in this analysis are required to have $p_T(\text{jet}) > 25\,\text{GeV}$ for central jets with $|\eta| \leq 2.4$, and $p_T(\text{jet}) > 30\,\text{GeV}$ for forward jets with $2.4 < |\eta| < 4.5$. Additional selection criteria, including cleaning cuts to remove fake jets and jets deposited through areas of the detector that are not functioning properly, are applied as detailed in Sect. 9.1.

b-Jets Identification of jets containing b-quarks, i.e., b-jets, is crucial for this analysis. In simulation, a jet is labeled at truth level as a b-jet if a b-quark (after final state radiation) with transverse momentum above $5\,\text{GeV}$ is identified within a cone of $\Delta R = 0.3$ around the jet axis. If no b-quark is identified, the jet is labeled as a charm quark jet if a charm quark is identified with the same criteria. If no charm quark is identified, the jet is labeled as a τ-jet if a τ-lepton is identified with the same criteria. Otherwise the jet is labeled as a light-flavor jet. The process of identifying b-jets is called b-tagging. In this thesis, unless otherwise specified, b-jets refer to reconstructed jets that are b-tagged.

Jets originating from b-quarks, which hadronize and decays, often present unique collider signatures. There will often be a secondary vertex where the B-hadron decays into lighter products, displaced from the primary vertex where the b-quark originates. The larger mass of B-hadrons relative to their decay products often leads to many more particles in the final state compared with light-quark jets, and the weak decay of B-hadrons can produce both charged leptons and neutrinos in the final state. All these unique properties are characterized and quantified as input variables to complicated algorithms, which decide the likelihood of a jet being a b-jet.

Jets considered for b-tagging have to pass the jet quality criteria, and have $p_T \geq 20$ GeV and $|\eta| < 2.5$. This analysis uses a b-tagging algorithm ("MV1") that has been optimized using multivariate techniques. The MV1 algorithm uses a large set of relevant variables as input, including the output variables of the other tagging algorithms: IP3D, a track-based algorithm, SV1 [39], a secondary vertex finding algorithm, and JetFitter, a neural network algorithm. These three input variables provide a continuous output value that is skewed towards one for b-jets and zero for non-b jets. The ATLAS flavor tagging working group provides several working points corresponding to canonical values of 60 %, 70 %, 80 % efficiency in identifying b-jets, and the 60 % efficiency working point (corresponding to an output value of above 0.9827 from the "MV1" b-tagging algorithm) is chosen in this analysis to maximize signal sensitivity.

The b-tagging efficiencies have been determined by the flavor tagging working group using the $t\bar{t}$ PDF method, which has higher precision compared with other existing methods. This method uses $t\bar{t}$ dilepton samples that have high purity and

8.3 Final-State Observables: Definition and Selection

are highly enriched in b-jets. The b-tagging efficiency can be extracted from a single equation for the fraction of tagged leading jets in data:

$$f_{\text{tagged}} = f_b \epsilon_b + (1 - f_b)\epsilon_l \tag{8.3}$$

where f_b is the fraction of real b-jets in the selected jet sample and ϵ_l is the light jet selection efficiency taken from simulation.

A scale factor (SF) then is defined as the ratio between the efficiency in data and that in simulation, for b-jets, c-jets, and light jets.

$$\text{SF}_b(p_\text{T}) = \frac{\epsilon_b^{\text{data}}}{\epsilon_b^{\text{MC}}}, \quad \text{SF}_c(p_\text{T}) = \frac{\epsilon_c^{\text{data}}}{\epsilon_c^{\text{MC}}}, \quad \text{SF}_l(p_\text{T}) = \frac{\epsilon_l^{\text{data}}}{\epsilon_l^{\text{MC}}} \tag{8.4}$$

The b-tagging performance is dependent on the transverse momentum of the jets. The b-tagging calibrations for b and c-jets only extend to 300 GeV in jet p_T with the light jets extending to 750 GeV. A MC-based analysis is used to assess an extrapolation uncertainty on the b-tagging efficiency from the last calibrated p_T bin up to 1200 GeV, to judge how systematic effects could impact the higher p_T jets compared to the last measured calibrated bin. The calculated scale factors are then used to determine a weight value to apply to each jet in an event with $p_\text{T} > 20$ GeV and $|\eta| < 2.5$ and subsequently a weight for the entire event. These weights correct the tagging rate in MC to that in data. They are obtained for the individual jets in two distinct ways. If the jet is tagged the weight is given by

$$w_{\text{jet}} = \text{SF}_{\text{flavour}}(p_\text{T}) \tag{8.5}$$

If the jet is not tagged the weight is calculated as:

$$w_{\text{jet}} = \frac{1 - \epsilon_{\text{flavour}}^{\text{data}}(p_\text{T})}{1 - \epsilon_{\text{flavour}}^{\text{MC}}(p_\text{T})} = \frac{1 - \text{SF}_{\text{flavour}}(p_\text{T})\epsilon_{\text{flavour}}^{\text{MC}}(p_\text{T})}{1 - \epsilon_{\text{flavour}}^{\text{MC}}(p_\text{T})} \tag{8.6}$$

The event weight applied is then the product of all the weights of the individual jets in that event.

$$w_{\text{event}} = \Pi_{\text{jet}} w_{\text{jet}} \tag{8.7}$$

The event weight is included in all the MC estimations after application of b-tagging. The scale factors are determined independently for b-jets, c-jets, and light jets and their uncertainties are uncorrelated. Therefore the b-tagging uncertainty is calculated separately for each kind of jet and the final systematic uncertainty due to the b-tagging is obtained by summing these three uncertainties in quadrature.

Electrons Electrons are preselected using the "Medium++" definition [40] with $p_\text{T} > 7$ GeV and $|\eta^{\text{clust}}| < 2.47$, where η is the pseudorapidity of the track if the track contains at least four silicon hits and that of the cluster otherwise. Electrons reconstructed in any of the problematic calorimeter regions (including the dead region present in the barrel EM calorimeter) are rejected both in data and MC. Smearing factors are applied to reconstructed electrons in MC events in order to more accurately reproduce the energy resolution of the calorimeter. In addition to the preselection, signal electrons used in the 1-lepton control regions are required to be pass the "Tight++" requirement [40], with $E^{\text{clust}}/\cosh\eta > 20\,\text{GeV}$, in order to meet the plateau of the single electron trigger, and to be well isolated with a cone size of $R = 0.3$. Scale factors that correct for discrepancies between data and MC are applied to all MC events with selected electrons.

Photons Although photons are ignored for most of the analysis, the data-driven $Z \to \nu\nu$ background estimation makes use of high p_T γ rays. Therefore only isolated photons with high p_T are reconstructed as photons. All others will either be reconstructed as jets, or only be considered in the E_T^{miss} calculation.

Photons are preselected using the 2012 tight definition [41] with $p_T > 10\,\text{GeV}$. Furthermore, only photons with $|\eta^{\text{clust}}| < 2.37$ are kept. Photons reconstructed in any of the problematic calorimeter regions (including the dead region present in the barrel EM calorimeter) are rejected both in data and MC. Smearing factors are applied to reconstructed photons in MC events in order to more accurately reproduce the energy resolution of the calorimeter.

In addition to the preselection, signal photons used in the $Z \to \nu\nu$ estimate are required to satisfy the isolation criteria $E^{\text{cone40}} < 5\,\text{GeV}$, in order to remove fake photon backgrounds, and have $p_T > 125\,\text{GeV}$ to meet the plateau of the single photon trigger (g120_loose).

Muons Muons are reconstructed using the "STACO" algorithm, which combines the inner detector and the muon spectrometer information (combined muons) [42]. To recover efficiency in the regions $|\eta| \approx 0$ and $|\eta| \approx 1.2$, segment tagged muons are also used. Muons are preselected for the analysis only if they have $p_T > 6\,\text{GeV}$ and $|\eta| < 2.5$. Furthermore, the following quality cuts on the tracks are applied:

- At least one hit in the B-layer if expected;
- At least one hit in any pixel layer;
- At least six SCT hits;
- The sum of the holes in the pixel and the SCT should be less than 3;
- A successful TRT-Extension where expected (i.e., within the acceptance of the TRT). An unsuccessful extension corresponds to either no TRT hit associated, or a set of TRT hits associated as outliers, i.e., outside the acceptance of the TRT. For $|\eta| < 1.9$, muons are required to have $N_{\text{TRT}} = N_{\text{TRT}}^{\text{hits}} + N_{\text{TRT}}^{\text{outliers}} > 5$. For $|\eta| > 1.9$, tracks with $N > 5$ should satisfy $N_{\text{TRT}}^{\text{outliers}} < 0.9 \times N_{\text{TRT}}$. $N_{\text{TRT}}^{\text{hits}}$ is the number of hits in the TRT associated with the track and $N_{\text{TRT}}^{\text{outliers}}$ is the number of TRT outliers on the muon track.

Furthermore, to avoid using cosmic muons, events containing a muon that survives overlap removal and has a transverse (longitudinal) impact parameter with respect to the primary vertex larger than 0.2 (1) mm are rejected. To correct for discrepancies in muon p_T resolution between data and MC, an additional smearing of the muon p_T is applied to simulated muons. In addition to the preselection, signal muons used in the one-lepton channel are required to have a transverse momentum larger than 20 GeV and be well isolated with $\Delta R < 0.4$.

E_T^{miss} The missing transverse momentum, the magnitude of which being the missing transverse energy E_T^{miss}, indicates the presence of neutrinos and possibly WIMP candidates in the final state. It is calculated with an algorithm based on a specific variant of E_T^{miss} [43, 44] called Egamma10NoTau as recommended, which consists of the sum of terms obtained, respectively, from the negative vector sum of the transverse momenta of jets, muons, electrons, photons, and topological calorimeter clusters not assigned to any reconstructed objects. The E_T^{miss} is calculated as follows:

$$\not{E}_T = \not{E}_T^e + \not{E}_T^\gamma + \not{E}_T^\mu + \not{E}_T^{\text{jets}} + \not{E}_T^{\text{SoftTerm}} \tag{8.8}$$

This order reflects the sequence of the objects used to calculate E_T^{miss} and the importance in defining the sequence for the signal ambiguity resolution: constituents used by a physics object contributing to E_T^{miss} are not used anymore in other contributions to E_T^{miss} to avoid energy double counting. Jets with $p_T > 20$ GeV and covering the full η range are included at the electromagnetic jet energy scale. The muon term includes all preselected muons before the overlap removal. Contribution from electrons includes electrons passing the "medium" electron selection criteria, with $p_T > 20$ GeV and before the overlap removal. The last term E_T^{SoftTerm} is calculated at the electromagnetic scale from topological calorimeter clusters that are not included in any reconstructed object. Details regarding the E_T^{miss} performance and reconstruction can be found in [44].

Overlap Removal According to the definitions above, energy deposits and tracks in the detector may be associated with multiple physics objects, and a procedure is implemented to remove duplication. The overlap removal procedure is carried out sequentially as follows:

- If an electron is found within $\Delta R < 0.05$ of another electron, remove the one with lower p_T.
- If a jet that is not b-tagged is found within $\Delta R < 0.2$ of an electron or photon passing the "loose" selection requirements, the jet is removed and the electron or photon is kept.
- After the previous step, if an electron, muon, or photon is found within $\Delta R < 0.4$ of a jet, the jet is kept and the electron, muon, or photon is removed.

References

1. T. Sjöstrand, S. Mrenna, P.Z. Skands, A brief introduction to PYTHIA 8.1. Comput. Phys. Commun. **178**, 852–867 (2008)
2. ATLAS Collaboration, The ATLAS Simulation Infrastructure. Eur. Phys. J. **C70**, 823–874 (2010)

3. S. Agostinelli et al., GEANT4: a simulation toolkit. Nucl. Instrum. Meth. **A506**, 250–303 (2003)
4. ATLAS Collaboration, The simulation principle and performance of the ATLAS fast calorimeter simulation FastCaloSim. (ATL-PHYS-PUB-2010-013) (2010)
5. T. Gleisberg, S. Hoeche, F. Krauss, M. Schonherr, S. Schumann, F. Siegert, J. Winter, Event generation with SHERPA 1.1. J. High Energy Phys. **2**, 007 (2009)
6. H.-L. Lai, M. Guzzi, J. Huston, Z. Li, P.M. Nadolsky, J. Pumplin, C.P. Yuan, New parton distributions for collider physics. Phys. Rev. **D82**, 074024 (2010)
7. S. Alioli, P. Nason, C. Oleari, E. Re, A general framework for implementing NLO calculations in shower Monte Carlo programs: the POWHEG BOX. J. High Energy Phys. **06**, 043 (2010)
8. P.Z. Skands, Tuning Monte Carlo generators: the Perugia tunes. Phys. Rev. **D82**, 074018 (2010)
9. S. Frixione, B.R. Webber, The MC@NLO 3.3 Event generator (2006)
10. S. Frixione, B.R. Webber, Matching NLO QCD computations and parton shower simulations. J. High Energy Phys. **06**, 029 (2002)
11. S. Frixione, E. Laenen, P. Motylinski, B.R. Webber, Single-top production in MC@NLO. J. High Energy Phys. **03**, 092 (2006)
12. J.M. Butterworth, J.R. Forshaw, M.H. Seymour, Multiparton interactions in photoproduction at HERA. Z. Phys. **C72**, 637–646 (1996)
13. B.P. Kersevan et al., The Monte Carlo event generator AcerMC versions 2.0 to 3.8 with interfaces to PYTHIA 6.4, HERWIG 6.5 and ARIADNE 4.1. Comput. Phys. Commun. **184**, 919–985 (2013)
14. S. Frixione, E. Laenen, P. Motylinski, B.R. Webber, C.D. White, Single-top hadroproduction in association with a W boson. J. High Energy Phys. **7**, 029 (2008)
15. M. Cacciari, M. Czakon, M. Mangano, A. Mitov, P. Nason, Top-pair production at hadron colliders with next-to-next-to-leading logarithmic soft-gluon resummation. Phys. Lett. **B710**, 612–622 (2012)
16. M. Beneke, P. Falgari, S. Klein, C. Schwinn, Hadronic top-quark pair production with NNLL threshold resummation. Nucl. Phys. **B855**, 695–741 (2012)
17. P. Baernreuther, M. Czakon, A. Mitov, Percent level precision physics at the Tevatron: first genuine NNLO QCD corrections to $q\bar{q} \to t\bar{t} + X$. Phys. Rev. Lett. **109**, 132001 (2012)
18. M. Czakon, A. Mitov, NNLO corrections to top pair production at hadron colliders: the quark-gluon reaction. J. High Energy Phys. **01**, 080 (2013)
19. M. Czakon, A. Mitov, NNLO corrections to top-pair production at hadron colliders: the all-fermionic scattering channels. J. High Energy Phys. **12**, 054 (2012)
20. M. Czakon, P. Fiedler, A. Mitov, The total top quark pair production cross-section at hadron colliders through $O(\alpha_S^4)$. Phys. Rev. Lett. **110**, 252004 (2013)
21. M. Czakon, A. Mitov, Top++: a program for the calculation of the top-pair cross-section at Hadron colliders. Comput. Phys. Commun. **185**, 2930 (2014)
22. M. Botje et al., The PDF4LHC working group interim recommendations. (2011)
23. A.D. Martin, W.J. Stirling, R.S. Thorne, G. Watt, Parton distributions for the LHC. Eur. Phys. J. **C63**, 189–285 (2009)
24. A.D. Martin, W.J. Stirling, R.S. Thorne, G. Watt, Uncertainties on alpha(S) in global PDF analyses and implications for predicted hadronic cross sections. Eur. Phys. J. **C64**, 653–680 (2009)
25. J. Gao, M. Guzzi, J. Huston, H.-L. Lai, Z. Li, P. Nadolsky, J. Pumplin, D. Stump, C.P. Yuan, CT10 next-to-next-to-leading order global analysis of QCD. Phys. Rev. **D89**, 033009 (2014)
26. R.D. Ball et al., Parton distributions with LHC data. Nucl. Phys. **B867**, 244–289 (2013)
27. G. Corcella, I.G. Knowles, G. Marchesini, S. Moretti, K. Odagiri, P. Richardson, M.H. Seymour, B.R. Webber, HERWIG 6: an event generator for hadron emission reactions with interfering gluons (including supersymmetric processes). J. High Energy Phys. **01**, 010 (2001)
28. J.M. Campbell, R.K. Ellis, MCFM for the tevatron and the LHC. Nucl. Phys. Proc. Suppl. **205–206**, 10–15 (2010)

29. T. Han S. Willenbrock, QCD correction to the p p —> W H and Z H total cross-sections. Phys. Lett. **B273**, 167–172 (1991)
30. O. Brein, A. Djouadi, R. Harlander, NNLO QCD corrections to the Higgs-Strahlung processes at hadron colliders. Phys. Lett. **B579**, 149–156 (2004)
31. M.L. Ciccolini, S. Dittmaier, M. Kramer, Electroweak radiative corrections to associated WH and ZH production at hadron colliders. Phys. Rev. **D68**, 073003 (2003)
32. D. Casadei et al., The implementation of the ATLAS missing Et triggers for the initial LHC operation. Technical Report ATL-DAQ-PUB-2011-00, CERN, Geneva (2011)
33. T. Barillari, E.B. Kuutmann, T. Carli, J. Erdmann, P. Giovannini, K.J. Grahn, C. Issever, A. Jantsch, A. Kiryunin, K. Lohwasser, A. Maslennikov, S. Menke, H. Oberlack, G. Pospelov, E. Rauter, P. Schacht, F. Spanó, P. Speckmayer, P. Stavina, P. Strízenec, Local hadronic calibration. Technical Report ATL-LARG-PUB-2009-001-2. ATL-COM-LARG-2008-006. ATL-LARG-PUB-2009-001, CERN, Geneva (2008). Due to a report-number conflict with another document, the report-number ATL-LARG-PUB-2009-001-2 has been assigned.
34. ATLAS Collaboration, Search for new phenomena in final states with an energetic jet and large missing transverse momentum in pp collisions at $\sqrt{s} = 8$ TeV with the ATLAS detector. Eur. Phys. J. **C75**, 299 (2015)
35. ATLAS Collaboration, Measurement of the muon reconstruction performance of the ATLAS detector using 2011 and 2012 LHC proton–proton collision data. Eur. Phys. J. **C74**, 3130 (2014)
36. W. Lampl, S. Laplace, D. Lelas, P. Loch, H. Ma, S. Menke, S. Rajagopalan, D. Rousseau, S. Snyder, G. Unal, Calorimeter clustering algorithms: description and performance. Technical Report ATL-LARG-PUB-2008-002. ATL-COM-LARG-2008-003, CERN, Geneva (2008)
37. Jet energy scale and its systematic uncertainty in proton-proton collisions at sqrt(s)=7 TeV in ATLAS 2010 data, Technical Report ATLAS-CONF-2011-032, CERN, Geneva (2011)
38. In-situ jet energy scale and jet shape corrections for multiple interactions in the first ATLAS data at the LHC, Technical Report ATLAS-CONF-2011-030, CERN, Geneva (2011)
39. Commissioning of the ATLAS high-performance b-tagging algorithms in the 7 TeV collision data, Technical Report ATLAS-CONF-2011-102, CERN, Geneva (2011)
40. C. Anastopoulos, E. Benhar-Noccioli, A. Bocci, K. Brendlinger, F. Bührer, L. Iconomidou-Fayard, M. Delmastro, O. Ducu, R. Fletcher, D. Froidevaux, T. Guillemin, S. Heim, F. Hubaut, M. Karnevskiy, J. Kretzschmar, J. Kroll, C. Lester, K. Lohwasser, J.B. Maurer, A. Morley, G. Pásztor, E. Richter-Was, A. Schaffer, T. Serre, P. Sommer, E. Tiouchichine, H. Williams, Supporting document on electron efficiency measurements using the 2012 LHC proton-proton collision data. Technical Report ATL-COM-PHYS-2013-1295, CERN, Geneva (2013)
41. ATLAS Collaboration, Electron and photon energy calibration with the ATLAS detector using LHC Run 1 data. Eur. Phys. J. **C74**, 3071 (2014)
42. ATLAS Collaboration, Preliminary results on the muon reconstruction efficiency, momentum resolution, and momentum scale in ATLAS 2012 pp collision data. Technical Report ATLAS-CONF-2013-088, CERN, Geneva (2013)
43. ATLAS Collaboration, Performance of missing transverse momentum reconstruction in proton-proton collisions at 7 TeV with ATLAS. Eur. Phys. J. **C72**, 1844 (2012)
44. Performance of Missing Transverse Momentum Reconstruction in ATLAS studied in Proton-Proton Collisions recorded in 2012 at 8 TeV. Technical Report ATLAS-CONF-2013-082, CERN, Geneva (2013)

Chapter 9
Dark Matter + Higgs($\rightarrow b\bar{b}$): Event Selection

This chapter presents a detailed description of the selection criteria used in this analysis to reject background and maximize signal sensitivity. The reconstructed objects used in the selection are jets, E_T^{miss}, electrons, and muons. A discussion of the definition of these objects can be found in Chap. 8, where the triggers used in this analysis are also discussed.

This chapter is organized as follows: Sect. 9.1 discusses the requirements placed on all signal and control region events, called "preselection" for high quality pp collision data to be used in the analysis. Section 9.2 describes the selection criteria that define the signal region, as well as the resulting selection efficiency of signal events.

9.1 Event Preselection

The selections discussed in this section are applied to events in all signal and control regions to ensure high quality data.

The events used in the analysis are recorded when all subdetectors are function correctly, as labeled belonging to the Good Run List (GRL). We are using the GRL from Feb. 2014 for 8 TeV collision data recorded in 2012, named `data12_8TeV.periodAllYear_DetStatus-v61-pro14-02 _DQDefects-00-01-00 _PHYS_StandardGRL_All_Good.xml`.

9.1.1 BCH Cleaning

For a variety of reasons, there are modules in the calorimeter which are either temporarily or permanently masked throughout all data taking periods. In 2012, there was one module which was off for the full dataset, and which was thus added to the MC as well. It turns out that the correction which was used to correct for masked cells within the tile calorimeter was not able to properly handle entire dead module(s), and thus some problems have been observed. In particular, high momentum jets are particularly sensitive, as they are more collimated and thus can be more completely contained within masked modules. Studies within the tile and jet groups have shown that medium to high momentum jets which fall within a masked module (this region is called "core") are usually under-corrected, while jets in modules adjacent to a masked module (this region is called "edge") are over-corrected. As the jet p_T is increased, this becomes a much more significant contribution. It becomes important to remove events where such jets fall into masked regions, as otherwise the jet is poorly reconstructed. This is also important for E_T^{miss} reconstruction, as these masked regions can create large amounts of fake E_T^{miss}. The latest version of the BCH Cleaning Tool [1] with the *medium* cleaning criteria is applied to all events used in this analysis. *Tight* cleaning has been tried as well: the change between *tight* and *medium* cleaning is small, and effectively disappears after b-tagging requirements, hence *medium* cleaning is adopted to retain selection efficiency of events. Table 9.1 shows the ratio of events observed before and after applying the requirement on the BCH cleaning for various background sources and data. The overall effect of BCH cleaning and the uncertainty associated with it is small.

9.1.2 Data Quality

Events are rejected for which the data quality requirements indicate possible problem with one or more of the ATLAS subdetectors.

9.1.3 Vertex Selection

To ensure that a recorded event is consistent with a pp-collision, events with less than five tracks associated with the selected primary vertex are rejected because we need high quality vertices for b-tagging. The minimum p_T required for a track associated with a vertex is 500 MeV.

Table 9.1 Ratio of event yield with and without the medium BCH requirement applied

	Diboson	V+jet	$t\bar{t}$	Multijet	$Z \to \nu\nu$	Exp Bkdg	Data
Ratio medium/no cut	1.00	1.00	1.00	0.97	0.97	0.98	0.98

9.1.4 Trigger

In the signal regions and 0-lepton control regions, events are selected by an E_T^{miss} trigger with a threshold of 80 GeV at the event filter (EF) level (`EF_xe80_tclcw`), which reaches an efficiency of 98 % at 151 GeV. For more details on the E_T^{miss} trigger and its efficiency, see Sect. 8.2.1.

Muon triggers with thresholds of 24 GeV for isolated muons, and 36 GeV for muons with no isolation requirement, are used to select muon data used for the validation of backgrounds in the 1-lepton control regions. A photon trigger with a threshold of 120 GeV is used to select events with a high p_T prompt photon for data-driven $Z \rightarrow \nu\nu$ background estimation. More details are given in Sect. 8.2.2.

9.1.5 Event Cleaning

Additional cleaning is applied to the remaining events. The following events are removed: events recorded in a time window around a noise burst in the LAr calorimeter, events recorded when there are functioning errors in the tile calorimeter or saturation in one of the tile calorimeter cells, and incomplete events where some detector information is missing.

9.1.6 Jet Cleaning

The main backgrounds to jets coming from pp-collisions are calorimeter noise and non-collision events like cosmic ray muons or beam-halo events. To reduce these contaminations, a number of quality requirements are placed on any jet with a calibrated p_T above 20 GeV within the full η range. If any of these jets does not pass the selection, the event is rejected. The quality requirements are detailed in [2] and correspond to the *looser* jet quality criterion, which was designed to provide an efficiency above 99.8 % while retaining a rejection of fake jets as high as possible, hence recommended for physics analysis.

An additional cleaning cut is used to further suppress beam backgrounds. Events are rejected if the leading jet does not satisfy chf $> 0.1 *$ fracSamplingMax, where chf is the jet charge fraction and "fracSamplingMax" the maximum energy fraction in one calorimeter layer.

9.1.7 Jet Vertex Fraction

The jet vertex fraction (JVF) is a quantity constructed from tracking and vertexing information, and used to identify and select jets originating in the hard-scatter interaction point and reject fake jets. It is calculated as the scalar sum of the track p_T originating from the primary vertex divided by the sum of all track p_T associated with the jet. Because of the limited η coverage of the ATLAS inner detectors that provide the tracking information, JVF can only be calculated for central jets. As jets not originating from the interaction point tend to have lower p_T, central jets with $p_T < 50$ GeV are required to have a JVF value of ≥ 0.5.

9.1.8 Jet Multiplicity

For the preselection we require the event to have at least one "good" jet, defined as passing the aforementioned jet reconstruction and selection requirements, with $p_T(\text{jet}) > 20$ GeV and $|\eta| < 2.8$.

9.1.9 E_T^{miss}

For the analysis of the full 2012 dataset, a variant of E_T^{miss} called Egamma10NoTau is used. We require any event selected for this analysis to have E_T^{miss} or modified E_T^{miss} (E_T^{miss} ll or E_T^{miss} g) above 100 GeV.

Details of the E_T^{miss} calculation are given in Sect. 8.3. Definitions of the modified E_T^{miss} are described in the $Z(\to \nu\bar{\nu}) + jets$ background estimation in Sect. 10.3.2.

9.1.10 Lepton Veto

In the signal regions and 0-lepton validation region, we reject events containing any electron or muon as defined in Sect. 8.3. In the lepton control regions, this cut is modified to require a specific number of reconstructed electrons and muons as described in Sect. 8.3.

9.2 Selection of $E_T^{\text{miss}} + h(\to b\bar{b})$ Signal

For the events passing the aforementioned "preselection" requirements, a dedicated set of selection criteria are developed and optimized to select signal events with $E_T^{\text{miss}} + h(\to b\bar{b})$ final state. The selections and the resulting efficiency for the Z'-2HDM signal events are described below.

9.2 Selection of $E_T^{miss} + h(\to b\bar{b})$ Signal

9.2.1 Signal Selection

The signal process, as shown in Fig. 7.1, leads to collider signatures with a pair of b-quarks in the final state from the Higgs boson decay recoiling against large missing transverse momentum from the WIMP pair. Due to the resonant production and decay of the heavy Z', the WIMP pair is boosted with small angular separation between the two dark matter particles. This leads to large missing transverse momentum in the final state even for low WIMP masses. The majority of signal samples produced, especially the ones with high $m_{Z'}$ relative to m_{A^0}, lead to E_T^{miss} peaks at very high values, motivating probes into high E_T^{miss} regions different from existing Standard Model (SM) Higgs searches in association with a vector boson.

This final state is very similar to the DM+$b(\bar{b})$ signal in the existing DM+heavy-flavor analysis [3] as mentioned in Chap. 5, where the selections are derived from. To select the DM + $h(\to b\bar{b})$ events, the variables providing the best separation between the SM backgrounds and signal events are found to be E_T^{miss}, leading b-jet p_T, sub-leading b-jet p_T, b-tagging requirement with a selected efficiency working point, $\Delta\phi_{min}(E_T^{miss}, j)$ (the smallest azimuthal angle between $\mathbf{E}_T^{miss}$ and jets), jet multiplicity (n_j), and requiring the invariant mass of the reconstructed two b-jets to be within the Higgs mass window. The optimal selection values are determined using an iterative process of maximizing signal significance "S" for the DM + $H(b\bar{b})$ signal, at different selection stages for different variables, where the signal significance S is defined as

$$S = \frac{N_S}{\sqrt{N_S + N_b + 0.2^2 \cdot N_b^2}} \quad (9.1)$$

where N_S is the expected number of signal events; N_b is the expected number of background events; and the third term in the denominator is a rough approximation of the systematic uncertainty which is conservatively estimated to be 20% of the number of background events.

The optimized signal selection criteria are listed as follows and applied in this progressive order to ensure good understanding of the background and data at each selection stage, where the symbol j represents an anti-k_t jet ($R = 0.4$), b a b-tagged anti-k_t jet ($R = 0.4$), and the subscript index i of each jet collection means the i-th jet in descending order of the transverse momentum, of which j_i are inclusive and may or may not be b-tagged.

- Lepton Veto
- $E_T^{miss} > 100\,\text{GeV}$
- $\Delta\phi_{min}(E_T^{miss}, p_T(j)) > 1.0$
- highest p_T jet must have $p_T > 100\,\text{GeV}$
- $\geq 1 b$-jet with b-tagging working point: 60%
- highest p_T b-jet must have $p_T > 100\,\text{GeV}$
- $2 \leq n_j \leq 3$

- sub-leading jet $p_T > 100\,\text{GeV}$ when $n_j = 3$
- sub-leading b-jet $p_T > 60\,\text{GeV}$ when $n_j = 3$ and $n_b > 1$
- $\geq 2b$-jet with b-tagging working point: 60 %
- $90\,\text{GeV} \leq m_{bb} \leq 150\,\text{GeV}$
- $\Delta R(p_T(j0), p_T(j1)) < 1.5$
- $E_T^{\text{miss}} > 150, 200, 300,$ or $400\,\text{GeV}$

To suppress contamination from multijet events, the smallest azimuthal angle between $\mathbf{E}_T^{\text{miss}}$ and small-R jets is required to be greater than 1.0. The additional requirements placed for events with high jet multiplicity, i.e., $n_j = 3$, are found to be effective in removing top quark background. Further taking advantage of the fact that the two b-jets coming from Higgs decay in the signal process have relatively small angular separation due to the high p_T of the Higgs, as seen in Figs. 7.6 and 7.7, an additional requirement of $\Delta R(p_T(j0), p_T(j1)) < 1.5$ is applied, which further suppresses background and avoids background mismodeling in b-enriched regions where the two b-jets coming from different vertices usually have a larger angular separation.

Finally, we use an increasing E_T^{miss} threshold to take full advantage of the hard E_T^{miss} spectrum for signal processes with large $m_{Z'}$ and small m_{A^0} in order to reach best signal sensitivity. A baseline $E_T^{\text{miss}} > 100\,\text{GeV}$ requirement is applied in the beginning, with the ascending E_T^{miss} threshold adopted at the end requiring $E_T^{\text{miss}} \geq 150, 200, 300,$ and $400\,\text{GeV}$. The best signal sensitivity at $\tan\beta = 1$ is achieved by requiring a minimum E_T^{miss} of $200\,\text{GeV}$ for $m_{Z'} = 600$, $300\,\text{GeV}$ for $m_{Z'} = 800$, and $400\,\text{GeV}$ for $m_{Z'} = 1000$–$1400\,\text{GeV}$.

9.2.2 Signal Selection Efficiency

We calculate the product of the detector acceptance and reconstruction efficiency (selection efficiency) in the aforementioned signal region for the $Z' \to A^0 h$ signal process in the ($m_{Z'}, m_{A^0}$) parameter space we probe. The results are given in Tables 9.2, 9.3, 9.4, 9.5, 9.6, and 9.7. We also calculate the selection efficiency for the $Z' \to Zh$ processes as shown in Table 9.8. In summary, the selection efficiency of the $Z' \to h(b\bar{b}) + E_T^{\text{miss}}$ signal after the full set of selection requirements varies from approximately 5–10 % depending on $m_{Z'}$ and m_A.

9.2 Selection of $E_T^{miss} + h(\to b\bar{b})$ Signal

Table 9.2 Exclusive and inclusive acceptance for $Z' \to A^0 h$ signal samples with $m_{A^0} = 300$ GeV, $\tan\beta = 1$, and different $m_{Z'}$

$m_{A^0} = 300$ GeV	$m_{Z'} = 600$ GeV		$m_{Z'} = 800$ GeV		$m_{Z'} = 1000$ GeV		$m_{Z'} = 1200$ GeV		$m_{Z'} = 1400$ GeV	
	Excl.	Incl.	Excl.	Incl.	Excl.	Incl.	Excl.	Incl.	Excl.	Incl.
$1 \leq n_j \leq 4$	0.96	0.96	0.95	0.95	0.94	0.94	0.94	0.94	0.95	0.95
$\Delta\phi_{min}(j, E_T^{miss}) > 1$	0.84	0.70	0.98	0.79	0.99	0.79	1.00	0.78	1.00	0.77
$p_T(j_0) > 100$ GeV	0.84	0.70	0.97	0.78	0.99	0.79	0.99	0.77	0.99	0.77
At least one b-btag 60 % eff.	0.73	0.51	0.75	0.58	0.74	0.58	0.71	0.55	0.59	0.46
$(n_j \geq 3) p_T(j_1) > 100$ GeV	0.74	0.38	0.79	0.46	0.84	0.49	0.85	0.47	0.85	0.39
$(n_j \geq 3) p_T(b_1) > 60$ GeV	0.99	0.38	0.98	0.46	0.99	0.48	0.99	0.47	0.97	0.38
$p_T(b_0) > 100$ GeV	0.81	0.30	0.84	0.38	0.85	0.41	0.83	0.39	0.81	0.32
$2 \leq n_j \leq 3$	0.72	0.22	0.83	0.32	0.85	0.35	0.86	0.34	0.74	0.23
At least two b-btag 60 % eff.	0.43	0.10	0.45	0.14	0.42	0.15	0.39	0.13	0.38	0.09
90 GeV $\leq m_{bb} \leq 150$ GeV	0.91	0.09	0.89	0.13	0.89	0.13	0.91	0.12	0.91	0.08
$\Delta R i(j_0, j_1) < 1.5$	0.81	0.07	0.98	0.12	0.98	0.13	0.96	0.12	0.90	0.08
$E_T^{miss} \geq 150$ GeV	0.92	0.07	1.00	0.12	1.00	0.13	1.00	0.12	1.00	0.08
$E_T^{miss} \geq 200$ GeV	0.33	0.02	0.96	0.12	0.99	0.13	0.99	0.12	1.00	0.08
$E_T^{miss} \geq 300$ GeV	0.04	0.00	0.46	0.05	0.88	0.11	0.96	0.12	0.97	0.08
$E_T^{miss} \geq 400$ GeV	0.16	0.00	0.02	0.00	0.51	0.06	0.85	0.10	0.91	0.07

Table 9.3 Exclusive and inclusive acceptance for $Z' \to A^0 h$ signal samples, $m_{A^0} = 400\,\text{GeV}$, $\tan\beta = 1$, and different $m_{Z'}$

$m_{A^0} = 400\,\text{GeV}$	$m_{Z'} = 600\,\text{GeV}$		$m_{Z'} = 800\,\text{GeV}$		$m_{Z'} = 1000\,\text{GeV}$		$m_{Z'} = 1200\,\text{GeV}$		$m_{Z'} = 1400\,\text{GeV}$	
	Excl.	Incl.	Excl.	Incl.	Excl.	Incl.	Excl.	Incl.	Excl.	Incl.
$1 \leq n_j \leq 4$	0.95	0.95	0.94	0.94	0.94	0.94	0.93	0.93	0.93	0.93
$\Delta\phi_{\min}(j, E_T^{\text{miss}}) > 1$	0.88	0.84	0.86	0.81	0.84	0.79	0.83	0.78	0.82	0.77
$p_T(j_0) > 100\,\text{GeV}$	0.69	0.58	0.95	0.77	0.99	0.78	0.99	0.78	0.99	0.77
At least one b-btag 60 % eff.	0.72	0.42	0.75	0.58	0.75	0.59	0.71	0.56	0.62	0.48
$(n_j \geq 3)\,p_T(j_1) > 100\,\text{GeV}$	0.76	0.32	0.76	0.45	0.82	0.49	0.85	0.48	0.84	0.42
$(n_j \geq 3)\,p_T(b_1) > 60\,\text{GeV}$	0.98	0.31	0.98	0.44	0.98	0.48	0.99	0.48	0.98	0.42
$p_t(b_0) > 100\,\text{GeV}$	0.84	0.27	0.81	0.36	0.85	0.41	0.85	0.41	0.80	0.34
$2 \leq n_j \leq 3$	0.65	0.17	0.81	0.29	0.84	0.35	0.86	0.35	0.75	0.26
At least two b-btag 60 % eff.	0.34	0.06	0.46	0.13	0.42	0.14	0.39	0.14	0.36	0.10
$90\,\text{GeV} \leq m_{bb} \leq 150\,\text{GeV}$	0.86	0.05	0.87	0.12	0.88	0.13	0.85	0.12	0.79	0.08
$\Delta R(j_0, j_1) < 1.5$	0.42	0.02	0.95	0.11	0.98	0.13	0.94	0.12	0.86	0.08
$E_T^{\text{miss}} \geq 150\,\text{GeV}$	0.76	0.02	0.99	0.11	1.00	0.13	1.00	0.12	1.00	0.09
$E_T^{\text{miss}} \geq 200\,\text{GeV}$	0.44	0.01	0.88	0.10	0.98	0.13	0.99	0.12	1.00	0.09
$E_T^{\text{miss}} \geq 300\,\text{GeV}$	0.24	0.00	0.15	0.01	0.85	0.11	0.95	0.12	0.97	0.09
$E_T^{\text{miss}} \geq 400\,\text{GeV}$	0.32	0.00	0.04	0.00	0.27	0.03	0.78	0.09	0.88	0.08

9.2 Selection of $E_T^{miss} + h(\to b\bar{b})$ Signal

Table 9.4 Exclusive and inclusive acceptance for $Z' \to A^0 h$ signal samples, $m_{A^0} = 500$ GeV, $\tan\beta = 1$, and different $m_{Z'}$

$m_{A^0} = 500$ GeV	$m_{Z'} = 800$ GeV		$m_{Z'} = 1000$ GeV		$m_{Z'} = 1200$ GeV		$m_{Z'} = 1400$ GeV	
	Excl.	Incl.	Excl.	Incl.	Excl.	Incl.	Excl.	Incl.
$1 \leq n_j \leq 4$	0.94	0.94	0.94	0.94	0.93	0.93	0.93	0.93
$\Delta\phi_{min}(j, E_T^{miss}) > 1$	0.87	0.82	0.85	0.80	0.83	0.78	0.82	0.78
$p_T(j_0) > 100$ GeV	0.89	0.73	0.98	0.79	0.99	0.78	0.97	0.75
At least one b-btag 60 % eff.	0.75	0.55	0.75	0.59	0.72	0.56	0.66	0.51
$(n_j \geq 3)\, p_T(j_1) > 100$ GeV	0.74	0.41	0.80	0.47	0.83	0.48	0.84	0.44
$(n_j \geq 3)\, p_T(b_1) > 60$ GeV	0.97	0.40	0.98	0.47	0.98	0.48	0.98	0.44
$p_t(b_0) > 100$ GeV	0.81	0.33	0.84	0.40	0.84	0.41	0.79	0.35
$2 \leq n_j \leq 3$	0.75	0.25	0.82	0.33	0.85	0.35	0.78	0.29
At least two b-btag 60 % eff.	0.42	0.10	0.42	0.14	0.38	0.13	0.36	0.11
90 GeV $\leq m_{bb} \leq 150$ GeV	0.90	0.10	0.86	0.13	0.84	0.12	0.75	0.09
$\Delta Ri(j_0, j_1) < 1.5$	0.87	0.09	0.96	0.13	0.92	0.12	0.81	0.09
$E_T^{miss} \geq 150$ GeV	0.95	0.09	1.00	0.13	1.00	0.13	1.00	0.11
$E_T^{miss} \geq 200$ GeV	0.63	0.05	0.97	0.12	0.99	0.13	1.00	0.11
$E_T^{miss} \geq 300$ GeV	0.07	0.00	0.67	0.08	0.92	0.12	0.97	0.10
$E_T^{miss} \geq 400$ GeV	0.12	0.00	0.09	0.01	0.71	0.08	0.88	0.09

Table 9.5 Exclusive and inclusive acceptance for $Z' \to A^0 h$ signal samples, $m_{A^0} = 600\,\text{GeV}$, $\tan\beta = 1$, and different $m_{Z'}$

$m_{A^0} = 600\,\text{GeV}$	$m_{Z'} = 800\,\text{GeV}$		$m_{Z'} = 1000\,\text{GeV}$		$m_{Z'} = 1200\,\text{GeV}$		$m_{Z'} = 1400\,\text{GeV}$	
	Excl.	Incl.	Excl.	Incl.	Excl.	Incl.	Excl.	Incl.
$1 \leq n_j \leq 4$	0.92	0.92	0.93	0.93	0.92	0.92	0.93	0.93
$\Delta\phi_{\min}(j, E_T^{\text{miss}}) > 1$	0.85	0.80	0.85	0.80	0.84	0.78	0.81	0.76
$p_T(j_0) > 100\,\text{GeV}$	0.76	0.63	0.96	0.77	0.97	0.77	0.98	0.76
At least one b-btag 60 % eff.	0.75	0.49	0.75	0.58	0.74	0.58	0.68	0.52
$(n_j \geq 3)\, p_T(j_1) > 100\,\text{GeV}$	0.70	0.34	0.78	0.45	0.81	0.47	0.80	0.44
$(n_j \geq 3)\, p_T(b_1) > 60\,\text{GeV}$	0.94	0.34	0.97	0.44	0.97	0.47	0.94	0.44
$p_t(b_0) > 100\,\text{GeV}$	0.79	0.29	0.81	0.36	0.84	0.40	0.82	0.38
$2 \leq n_j \leq 3$	0.67	0.20	0.80	0.30	0.81	0.33	0.83	0.32
At least two b-btag 60 % eff.	0.40	0.09	0.43	0.13	0.40	0.14	0.35	0.12
$90\,\text{GeV} \leq m_{bb} \leq 150\,\text{GeV}$	0.66	0.06	0.84	0.12	0.82	0.12	0.88	0.12
$\Delta R(j_0, j_1) < 1.5$	0.73	0.06	0.93	0.12	0.91	0.12	0.96	0.12
$E_T^{\text{miss}} \geq 150\,\text{GeV}$	0.89	0.05	0.99	0.12	1.00	0.13	1.00	0.12
$E_T^{\text{miss}} \geq 200\,\text{GeV}$	0.59	0.03	0.93	0.11	0.99	0.13	1.00	0.12
$E_T^{\text{miss}} \geq 300\,\text{GeV}$	0.17	0.01	0.34	0.04	0.88	0.12	0.96	0.11
$E_T^{\text{miss}} \geq 400\,\text{GeV}$	0.19	0.00	0.08	0.00	0.51	0.06	0.86	0.10

9.2 Selection of $E_T^{miss} + h(\to b\bar{b})$ Signal

Table 9.6 Exclusive and inclusive acceptance for $Z' \to A^0 h$ signal samples, $m_{A^0} = 700$ GeV, $\tan\beta = 1$, and different $m_{Z'}$

$m_{A^0} = 700$ GeV	$m_{Z'} = 1000$ GeV		$m_{Z'} = 1200$ GeV		$m_{Z'} = 1400$ GeV	
	Excl.	Incl.	Excl.	Incl.	Excl.	Incl.
$1 \leq n_j \leq 4$	0.93	0.93	0.91	0.91	0.91	0.91
$\Delta\phi_{min}(j, E_T^{miss}) > 1$	0.86	0.81	0.82	0.77	0.80	0.75
$p_T(j_0) > 100$ GeV	0.91	0.75	0.98	0.77	0.96	0.75
At least one b-btag 60 % eff.	0.72	0.54	0.74	0.58	0.71	0.56
$(n_j \geq 3) \, p_T(j_1) > 100$ GeV	0.74	0.42	0.80	0.47	0.84	0.48
$(n_j \geq 3) \, p_T(b_1) > 60$ GeV	0.92	0.39	0.98	0.47	0.98	0.48
$p_t(b_0) > 100$ GeV	0.80	0.33	0.82	0.39	0.82	0.40
$2 \leq n_j \leq 3$	0.71	0.24	0.77	0.30	0.77	0.32
At least two b-btag 60 % eff.	0.34	0.09	0.42	0.14	0.34	0.12
90 GeV $\leq m_{bb} \leq 150$ GeV	0.77	0.09	0.77	0.11	0.81	0.12
$\Delta R(j_0, j_1) < 1.5$	0.86	0.09	0.86	0.11	0.90	0.12
$E_T^{miss} \geq 150$ GeV	0.98	0.09	1.00	0.13	1.00	0.13
$E_T^{miss} \geq 200$ GeV	0.76	0.07	0.98	0.12	0.99	0.13
$E_T^{miss} \geq 300$ GeV	0.22	0.02	0.77	0.09	0.95	0.12
$E_T^{miss} \geq 400$ GeV	0.12	0.00	0.23	0.02	0.77	0.09

Table 9.7 Exclusive and inclusive acceptance for $Z' \to A^0 h$ signal samples, $m_{A^0} = 800$ GeV, $\tan\beta = 1$, and different $m_{Z'}$

$m_{A^0} = 800$ GeV	$m_{Z'} = 1000$ GeV		$m_{Z'} = 1200$ GeV		$m_{Z'} = 1400$ GeV	
	Excl.	Incl.	Excl.	Incl.	Excl.	Incl.
$1 \leq n_j \leq 4$	0.94	0.94	0.93	0.93	0.93	0.93
$\Delta\phi_{min}(j, E_T^{miss}) > 1$	0.88	0.82	0.85	0.79	0.84	0.78
$p_T(j_0) > 100$ GeV	0.85	0.70	0.97	0.77	0.99	0.78
At least one b-btag 60 % eff.	0.76	0.53	0.76	0.59	0.73	0.57
$(n_j \geq 3) \, p_T(j_1) > 100$ GeV	0.75	0.40	0.78	0.46	0.84	0.48
$(n_j \geq 3) \, p_T(b_1) > 60$ GeV	0.98	0.39	0.99	0.45	0.99	0.48
$p_t(b_0) > 100$ GeV	0.82	0.32	0.84	0.38	0.85	0.41
$2 \leq n_j \leq 3$	0.76	0.25	0.82	0.31	0.84	0.34
At least two b-btag 60 % eff.	0.43	0.11	0.43	0.13	0.42	0.14
90 GeV $\leq m_{bb} \leq 150$ GeV	0.90	0.10	0.90	0.12	0.90	0.13
$\Delta R(j_0, j_1) < 1.5$	0.84	0.08	0.98	0.12	0.99	0.13
$E_T^{miss} \geq 150$ GeV	0.93	0.07	0.99	0.12	1.00	0.13
$E_T^{miss} \geq 200$ GeV	0.72	0.05	0.93	0.11	0.99	0.13
$E_T^{miss} \geq 300$ GeV	0.37	0.02	0.49	0.05	0.90	0.11
$E_T^{miss} \geq 400$ GeV	0.22	0.00	0.18	0.01	0.62	0.07

Table 9.8 Exclusive and inclusive acceptance for $Z' \to Zh$ process at different $m_{Z'}$ and $\tan\beta = 1$

$Z' \to Zh$	$m_{Z'} = 600\,\text{GeV}$		$m_{Z'} = 800\,\text{GeV}$		$m_{Z'} = 1000\,\text{GeV}$		$m_{Z'} = 1200\,\text{GeV}$		$m_{Z'} = 1400\,\text{GeV}$	
	Excl.	Incl.	Excl.	Incl.	Excl.	Incl.	Excl.	Incl.	Excl.	Incl.
$1 \leq n_j \leq 4$	0.96	0.96	0.95	0.95	0.94	0.94	0.94	0.94	0.94	0.94
$\Delta\phi_{\min}(j, E_T^{\text{miss}}) > 1$	0.86	0.82	0.85	0.81	0.84	0.79	0.82	0.78	0.80	0.76
$p_T(j_0) > 100\,\text{GeV}$	0.94	0.77	0.98	0.79	0.98	0.78	0.99	0.78	0.98	0.76
at least one b-btag 60 % eff.	0.74	0.57	0.74	0.58	0.72	0.56	0.67	0.53	0.55	0.42
$(n_j \geq 3)\, p_T(j_1) > 100\,\text{GeV}$	0.76	0.43	0.82	0.48	0.85	0.48	0.84	0.44	0.83	0.36
$(n_j \geq 3)\, p_T(b_1) > 60\,\text{GeV}$	0.99	0.43	0.99	0.47	0.99	0.47	0.97	0.44	0.98	0.36
$p_t(b_0) > 100\,\text{GeV}$	0.82	0.35	0.84	0.40	0.85	0.40	0.80	0.36	0.74	0.27
$2 \leq n_j \leq 3$	0.80	0.28	0.85	0.34	0.86	0.35	0.82	0.31	0.73	0.21
at least two b-btag 60 % eff.	0.46	0.13	0.41	0.14	0.39	0.14	0.38	0.12	0.28	0.06
$90\,\text{GeV} \leq m_{bb} \leq 150\,\text{GeV}$	0.89	0.11	0.89	0.13	0.89	0.13	0.87	0.11	0.83	0.06
$\Delta R(j_0, j_1) < 1.5$	0.94	0.11	0.97	0.12	0.97	0.13	0.95	0.11	0.90	0.06
$E_T^{\text{miss}} \geq 150\,\text{GeV}$	0.99	0.11	0.99	0.12	0.98	0.13	0.96	0.11	0.91	0.06
$E_T^{\text{miss}} \geq 200\,\text{GeV}$	0.88	0.09	0.96	0.12	0.98	0.13	0.96	0.11	0.91	0.06
$E_T^{\text{miss}} \geq 300\,\text{GeV}$	0.07	0.01	0.75	0.09	0.91	0.11	0.84	0.09	0.93	0.06
$E_T^{\text{miss}} \geq 400\,\text{GeV}$	0.00	0.00	0.10	0.01	0.70	0.08	0.88	0.09	0.89	0.06

References

1. C. Doglioni, Bchcleaningtool (2013). Online https://twiki.cern.ch/twiki/bin/viewauth/AtlasProtected/BCHCleaningTool
2. ATLAS Collaboration, Selection of jets produced in proton-proton collisions with the ATLAS detector using 2011 data. Technical Report ATLAS-CONF-2012-020, CERN, Geneva, March 2012
3. ATLAS Collaboration, Search for dark matter in events with heavy quarks and missing transverse momentum in pp collisions with the ATLAS detector. Eur. Phys. J. **C75**, 92 (2015)

Chapter 10
Dark Matter + Higgs($\to b\bar{b}$): Background Processes

This chapter gives a detailed description of the main background processes in this analysis, how they are estimated and the result of the modeling in various dedicated control regions and validation regions. The selections are placed on physics objects described in Chap. 8. The control regions and validation regions are defined to be similar but orthogonal to the signal region, which is described in Chap. 9.

This chapter is organized as follows: Sect. 10.1 gives an overview of the main background processes, whether they are estimated through MC simulation or data-driven methods, and sketches out the control regions used to examine the modeling versus data; the simulated electroweak and top quark backgrounds are described in Sect. 10.2; the data-driven estimations of the multijet and $Z(\to \nu\bar{\nu})$+jets background are given in Sects. 10.3.1 and 10.3.2, respectively; and finally, the overall modeling of all background processes combined as described in the zero-lepton validation region is given in Sect. 10.4.

10.1 Background Overview

The main background processes in this analysis are diboson, $W(\to \ell\nu)$ and $Z(\to \ell\ell)$ + jets, top quark (including $t\bar{t}$ and single top quark), multijet, and $Z(\to \nu\bar{\nu})$ + jets. We use both MC samples and data-driven methods to estimate the different backgrounds, and create dedicated control regions for each process to study the modeling and constrain the normalization.

The simulated background processes are diboson, $W(\to \ell\nu)$ & $Z(\to \ell\ell)$ + jets ("V+jets"), and top quark. The diboson samples are well-studied and widely used in existing analyses, with proven good description of the process, hence no separate validation of the diboson sample is done for this analysis. For V+jets and top, we create 1-lepton control regions nearly identical to the signal selection, but reversing the lepton veto to require exactly one tight lepton (electron or muon) in the final state. The simulated background processes are detailed in Sect. 10.2. Good agreement is achieved between data and background.

The multijet background is estimated using a data-driven method called "jet smearing." We selected multijet enriched control regions in low $\Delta\phi_{\min}(E_\mathrm{T}^\mathrm{miss}, p_\mathrm{T}(j))$ and low $E_\mathrm{T}^\mathrm{miss}$ for this study, as detailed in Sect. 10.3.1. Good agreement is achieved, and it is also worth noting that multijet has little contribution to our final signal region.

The main irreducible background is $Z(\to \nu\bar{\nu})$ + jets, which is estimated with a data-driven method as detailed in Sect. 10.3.2. $Z \to \nu\bar{\nu}$ is estimated from reweighted $Z \to \mu\mu$ events in the low $E_\mathrm{T}^\mathrm{miss}$ region below 200 GeV, and from reweighted γ +jets events in the high $E_\mathrm{T}^\mathrm{miss}$ region above 200 GeV. Dedicated control regions are created to select $Z \to \mu\mu$ and γ + jets events. The final combined $Z \to \nu\bar{\nu}$ estimation is examined in the $Z \to \nu\bar{\nu}$ control region, which is nearly identical to the signal selection, but reversing the $2b$ jet requirement to either a b-veto or exactly $1b$ jet, the latter further blinded by requiring $E_\mathrm{T}^\mathrm{miss} \leq 200\,\mathrm{GeV}$; the invariant mass requirement on the two leading b-jets in the signal region is replaced by placing the same requirement on the invariant mass of the two leading jets, i.e., $90\,\mathrm{GeV} \leq m_{jj} \leq 150\,\mathrm{GeV}$ in the $Z \to \nu\bar{\nu}$ control region.

Finally, after obtaining and validating the background processes in their respective control regions, all background processes are put together in the 0-lepton validation region, as detailed in Sect. 10.4, which has near identical selection as the signal region, but with the cut on the invariant mass of the two leading b-jets, m_{bb}, reversed to require $m_{bb} < 60\,\mathrm{GeV}$ or $m_{bb} > 150\,\mathrm{GeV}$.

All control regions and validation regions are orthogonal to the final signal region, and good agreement between data and background is observed, as will be shown in subsequent sections.

10.2 Simulated Background Processes

We use MC generated samples to estimate W +jets, $Z(\to \ell\ell)$ +jets, diboson, single top, and $t\bar{t}$ processes. We select various control regions to ensure correct modeling of the selected data samples using these simulations. Section 10.2.1 presents selections and validation for the W + jets processes. Section 10.2.2 does so for a kinematic region dominated by top quarks.

The MC simulations are normalized according to their production cross-section and the luminosity of the 2012 data taking period ($20.3\,\mathrm{fb}^{-1}$). Event corrections like pile-up re-weighting, lepton identification scale factors, and b-tagging scale factors are also applied.

10.2 Simulated Background Processes

Table 10.1 Selection of the $V + \text{jets}$ control region

Electron or muon trigger	
Lepton quality	Tight lepton
$p_T(\ell)$	$> 25\,\text{GeV}$
$n(\ell)$	1
E_T^{miss}	$> 100\,\text{GeV}$
Jet multiplicity	$n_{\text{jet}} = 2$
$\Delta\phi_{\min}(E_T^{\text{miss}}, p_T(j))$	> 1.0
$p_T(\text{leading jet})$	$> 100\,\text{GeV}$
b-tagging working point: 60 %	$\geq 1b$-jet
$p_T(\text{leading } b\text{-jet})$	$> 100\,\text{GeV}$
$\Delta R(p_T(j_0), p_T(j_1))$	< 2.0
b-tagging working point: 60 %	$\geq 2b$-jet

10.2.1 W + Jets Control Region

We create a dataset enriched in $W + \text{jets}$ events by applying the criteria shown in Table 10.1. As top quark processes usually generate events with higher jet multiplicity, ($n_{\text{jet}} = 3$ in this analysis), for the $V + \text{jets}$ control region we choose events with lower jet multiplicity ($n_{\text{jet}} = 2$), and apply the signal region selection in the full E_T^{miss} range above 100 GeV, with the lepton veto replaced by requesting exactly one lepton (muon or electron) that passed the quality requirements in the final state. The $V + \text{jets}$ events selected are predominantly $W(\to \ell\nu) + \text{jets}$; $Z(\to \ell\ell) + \text{jets}$ contribute about 5 % of the $V + \text{jets}$ events before requiring b-tagged jets in the final state, and 3 % or less after b-tagging. In the ensuing tables and plots, we combine these two processes and name it $V + \text{jets}$.

Results are shown requiring exactly one muon in the final state. The resulting event yield at different selection stages is given in Table 10.2. As $V + \text{jets}$ events decrease significantly and are no longer the dominate process compared to top quark events after requiring b-tagged jets in the final state, we use the selection stages before b-tagging to derive a flat scale factor of 0.92 by keeping the event counts of the other backgrounds constant, and normalizing $V+\text{jets}$ events to match data. The observation and scale factor derived are consistent with existing analysis like the search for SM Higgs boson in the $V + h(\to b\bar{b})$ channel. The deviation from unity in the scale factor, namely 8 %, is much smaller than the uncertainty on the cross-section for $W+\text{jets}$ process, so no additional uncertainty is assigned to the normalization process.

Comparison between data and background with the scale factor applied to the $V + \text{jets}$ events is performed for different kinematic distributions including jet p_T, angular distributions, multiplicity, and E_T^{miss}. Good agreement is achieved for various selections stages, including before b-tagging (Fig. 10.1) and requiring at least $1b$-tagged jet in the final state (Fig. 10.2).

Table 10.2 Event yield for the W + jets control region with exactly one muon in the final state

	Diboson	W/Z + jets	Top	Expected backgrounds	Data	Ratio
Tight muon, $n_j = 2$	5267 ± 23	$412{,}078 \pm 568$	$22{,}435 \pm 42$	$439{,}780 \pm 570$	450,907	0.98
$E_T^{miss} > 100\,\text{GeV}$	1410 ± 12	$63{,}100 \pm 146$	5195 ± 19	$69{,}705 \pm 147$	65,467	1.06
$\Delta\phi_{min} > 1.0$	1264 ± 11	$49{,}715 \pm 125$	3874 ± 17	$54{,}853 \pm 126$	50,641	1.08
$p_T(j^0) > 100\,\text{GeV}$	1263 ± 11	$49{,}705 \pm 125$	3868 ± 17	$54{,}836 \pm 126$	50,613	1.08
1 b-tag 60%	134 ± 4	1852 ± 19	2303 ± 12	4289 ± 23	4022	1.07
$p_T(b^0) > 100\,\text{GeV}$	94 ± 3	1055 ± 15	1410 ± 9	2559 ± 18	2467	1.04
$\Delta R(j_0, j_1) < 2.0$	81 ± 3	597 ± 9	851 ± 7	1529 ± 12	1430	1.07
At least two b-btag 60% eff.	21 ± 1	51 ± 2	190 ± 3	262 ± 4	253	1.04

The uncertainty is statistical only. Only events with two jets are considered

10.2 Simulated Background Processes

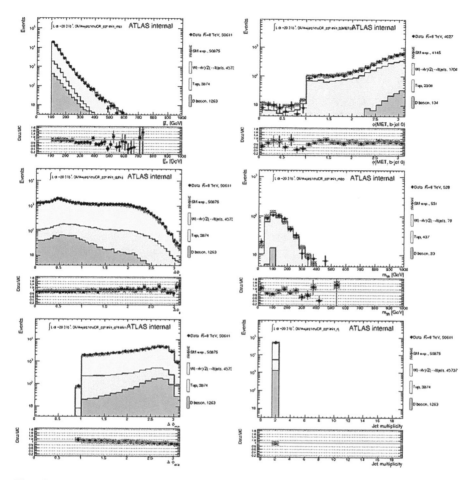

Fig. 10.1 Kinematic distributions in the $W+$ jets control region with one muon, before b-tagging requirement. Systematic uncertainties are given as hatched band and statistical uncertainties as error bars. Note as is in all subsequent plots, when the value m_{bb} is plotted, it is only for events with at least two b-jets. A scale factor of 0.92 is applied to the $W/Z+$ jets processes

10.2.2 Top Quark Control Region

In order to validate the simulated top quark background, including both $t\bar{t}$ and single top quark processes, we create a control region enriched in semileptonic decays of top quarks by applying the selection criteria shown in Table 10.3. This selection is identical to the signal region selection for events with high jet multiplicity ($n_{\text{jet}} = 3$) in the full E_T^{miss} range above 100 GeV, but with the lepton veto replaced by requesting exactly one lepton (muon or electron) that passed the quality requirements in the final state. The top quark events selected are predominantly $t\bar{t}$; single top quark

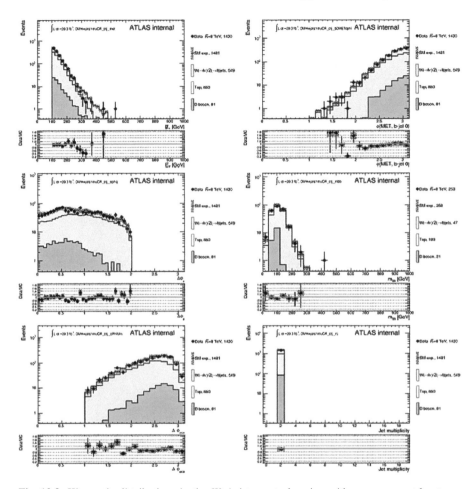

Fig. 10.2 Kinematic distributions in the W + jets control region with one muon, at least one b-tagged jet in the final state. Systematic uncertainties are given as hatched band and statistical uncertainties as error bars. Note as is in all subsequent plots, when the value m_{bb} is plotted, it is only for events with at least two b-jets. A scale factor of 0.92 is applied to the W/Z+jets processes

events contribute about 10 % of the total top quark event yield. In the ensuing tables and plots, we combine these two processes.

Results are shown for requiring exactly one muon in the final state. The resulting event yields at different selection stages are given in Table 10.4. Data and background comparison are performed for different kinematic distributions including jet p_T, angular distributions, multiplicity, and E_T^{miss}. The scale factor of 0.92 as derived from Sect. 10.2.1 is applied to the V+jets process already. Excellent agreement is achieved for various selections stages, including both before applying

10.2 Simulated Background Processes

Table 10.3 Selection of the top quark control region

Electron or muon trigger	
Lepton quality	Tight lepton
$p_T(\ell)$	$> 25\,\text{GeV}$
$n(\ell)$	1
E_T^{miss}	$> 100\,\text{GeV}$
Jet multiplicity	$n_{\text{jet}} = 3$
$\Delta\phi_{\text{min}}(E_T^{\text{miss}}, p_T(j))$	> 1.0
$p_T(\text{leading jet})$	$> 100\,\text{GeV}$
b-tagging working point: 60 %	$\geq 1b$-jet
$p_T(\text{leading } b\text{-jet})$	$> 100\,\text{GeV}$
$p_T(\text{sub-leading jet})$	$> 100\,\text{GeV}$
$p_T(\text{sub-leading } b\text{-jet})$ if $n_b > 1$	$> 60\,\text{GeV}$
$\Delta R(p_T(j_0), p_T(j_1))$	< 2.0
b-tagging working point: 60 %	$\geq 2b$-jet

the second b-tagged jet requirement, i.e., $\geq 1b$-tag (Fig. 10.3), and after, i.e., $\geq 2b$-tag (Fig. 10.4). Based on the good description, no additional scale factor correction is applied to the top events. As MC simulation predicts a harder p_T spectrum for $t\bar{t}$ events in comparison to data, a method called $t\bar{t}$ p_T reweighting is developed to correct MC prediction at the level of generated top quarks to match the distributions in the data [1, 2]. The effect of this correction is found to be 5.5 % and is small compared with the total systematic uncertainty on the $t\bar{t}$ process; hence it is not used for the results, but accounted for as an additional source of systematic uncertainty, as discussed in Sect. 11.

10.2.3 1-Lepton Validation Region

To examine the overall modeling of both $W/Z + $ jets and top quark processes, the 1-lepton validation region is defined by removing the jet multiplicity requirement separating the aforementioned $W + $ jets and top quark control regions. The purpose is to get a region with higher statistics, and selection requirements very close to those of the signal region with only the lepton veto reversed to require exactly one good muon in the final state. Figures 10.5 and 10.6 show various distributions in this combined 1-Lepton validation region both before and after the second b-tagged jet requirement. The scale factor of 0.92 as derived from Sect. 10.2.1 is applied to $W/Z + $ jets process. Good agreement is achieved for different kinematic properties at each selection stage.

Table 10.4 Event yield for the top control region with exactly one muon in the final state

	Diboson	V + jets	Top	Expected backgrounds	Data	Ratio
Tight muon, $n_j = 3$	3256 ± 18	$200{,}430 \pm 403$	$48{,}337 \pm 55$	$252{,}023 \pm 407$	266,990	0.94
$E_T^{miss} > 100\,\text{GeV}$	732 ± 8	$28{,}794 \pm 89$	9961 ± 25	$39{,}487 \pm 93$	38,128	1.04
$\Delta\phi_{min} > 1.0$	482 ± 7	$16{,}851 \pm 63$	5403 ± 18	$22{,}736 \pm 66$	21,947	1.04
$p_T(j^0) > 100\,\text{GeV}$	482 ± 7	$16{,}849 \pm 63$	5399 ± 18	$22{,}730 \pm 66$	21,937	1.04
1 b-tag 60 %	61 ± 2	1036 ± 11	3686 ± 14	4783 ± 18	4569	1.05
$p_T(b^0) > 100\,\text{GeV}$	36 ± 2	481 ± 8	1882 ± 10	2399 ± 13	2366	1.01
$p_T(j^1) > 100\,\text{GeV}$	11 ± 1	191 ± 5	496 ± 5	698 ± 7	699	1.00
$p_T(b^1) > 100\,\text{GeV}$	10 ± 1	184 ± 5	428 ± 5	622 ± 7	632	0.98
$\Delta R(j_0, j_1) < 2.0$	6 ± 1	67 ± 2	212 ± 3	285 ± 4	272	1.05
At least two b-btag 60 % eff.	1 ± 0	4 ± 0	32 ± 1	38 ± 1	40	0.95

The uncertainty is statistical. Only events with three jets are considered

10.3 Data-Driven Background Processes

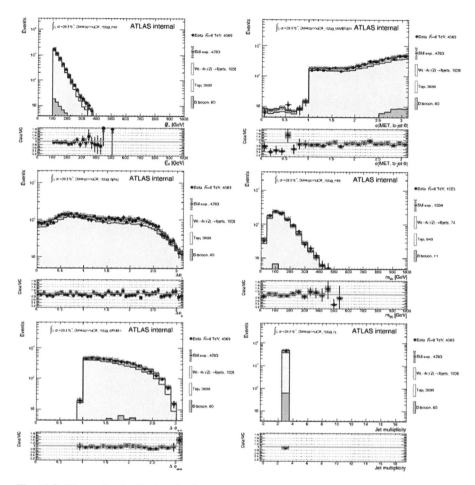

Fig. 10.3 Kinematic distributions in the top quark control region with one muon, at least one b-tagged jet in the final state. Systematic uncertainties are given as hatched band and statistical uncertainties as error bars. Note as is in all subsequent plots, when the value m_{bb} is plotted, it is only for events with at least two b-jets. A scale factor of 0.92 is applied to the W/Z+jets processes

10.3 Data-Driven Background Processes

Data-driven methods are used to estimate multijet background as detailed in Sect. 10.3.1, and the main irreducible background in this analysis, $Z(\rightarrow \nu\bar{\nu})$+jets as detailed in Sect. 10.3.2.

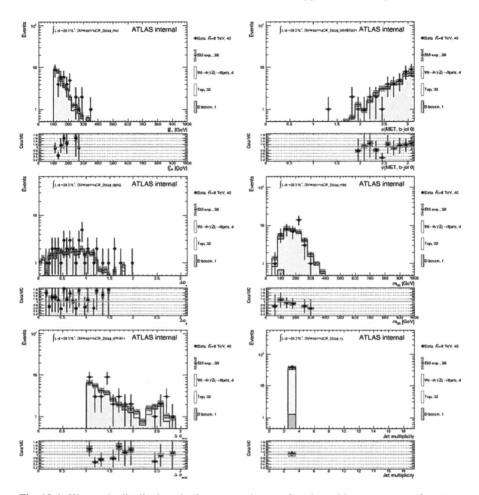

Fig. 10.4 Kinematic distributions in the top quark control region with one muon, at least two b-tagged jets in the final state. Systematic uncertainties are given as hatched band and statistical uncertainties as error bars. Note as is in all subsequent plots, when the value m_{bb} is plotted, it is only for events with at least two b-jets. A scale factor of 0.92 is applied to the $W/Z+$jets processes

10.3.1 Estimation of Multijet Background

Multijet events from high energy pp collisions generally have a large production cross-section, but rarely do they present large E_T^{miss}. This makes using MC simulation to estimate multijet events difficult. Therefore, we estimate the multijet background using a data-driven method called "jet smearing."

The smearing is performed using the "jet response function," which quantifies the probability of fluctuations in the detector to jets measured in the data. In the case

10.3 Data-Driven Background Processes

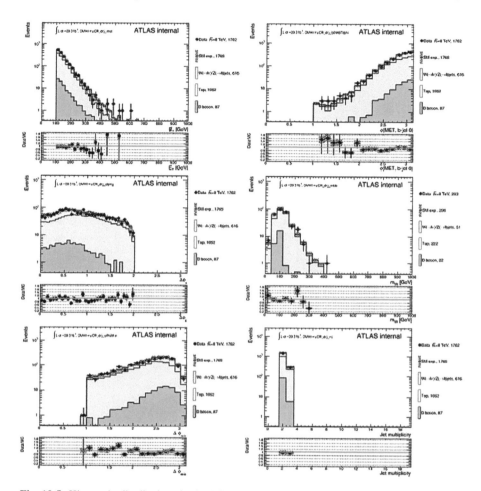

Fig. 10.5 Kinematic distributions in the 1-lepton validation region with one muon, at least one b-tagged jet in the final state. Systematic uncertainties are given as hatched band and statistical uncertainties as error bars. Note as is in all subsequent plots, when the value m_{bb} is plotted, it is only for events with at least two b-jets. A scale factor of 0.92 is applied to the W/Z+jets processes

of p_T response, the response distribution is defined as the p_T of the reconstructed jet divided by the p_T of the jet at truth level. Due to characteristics of jets and related detector performance, the response distribution can be broad as compared to a purely electro-magnetic object. The response function is constructed starting from multijet MC samples and modified to match data, and a combination of Gaussian and non-Gaussian functions are used to correctly account for the shape, including

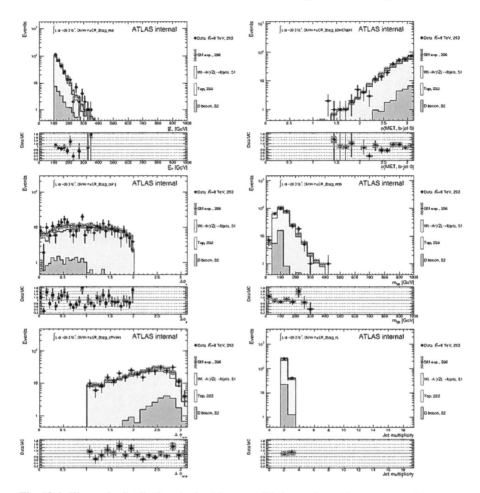

Fig. 10.6 Kinematic distributions in the 1-lepton validation region with one muon, at least two b-tagged jets in the final state. Systematic uncertainties are given as hatched band and statistical uncertainties as error bars. Note as is in all subsequent plots, when the value m_{bb} is plotted, it is only for events with at least two b-jets. A scale factor of 0.92 is applied to the W/Z+jets processes

the non-Gaussian tails of the response distribution. The jet smearing method, and the response function used, are well established by existing analyses and a detailed description can be found in [3].

Based on this method, to estimate multijet background, first, a selection of seed events is chosen such that they are highly enriched in typical, low E_T^{miss} multijet events. The jet response function, measured using a combination of data and Monte Carlo, is then used to smear the energy of each jet in the event, simulating the energy resolution of the calorimeters. Additional ϕ smearing is also performed in order to

10.3 Data-Driven Background Processes

model the ϕ resolution of the calorimeters. Each seed event can be duplicated and smeared an arbitrary number of times to achieve adequate statistics in the signal region. The resulting smeared "pseudo-data" acts as an effective simulation of the multijet background in any 0-lepton analysis.

Extra care is needed to make sure that this smeared pseudo-data well-models the multijet background. If there are too many smeared events, there will be double counting and the errors on the background will become misleadingly small. On the other hand, if too few seed events are used the resulting pseudo-data can have unwanted structure that shows up as peaks corresponding to the original seed events.

Finally, it is worth noting that multijet events have little to no contribution in the final signal region, as seen in Table 12.1. Some statistical effect is observed at earlier selection stages associated with the number of times the seed events are smeared, but it has no impact on the final signal region where multijet events have been effectively removed.

10.3.1.1 Seed Selection

For our seed selection the following requirements are made to select multijet enriched events from `JetTauEtmiss` stream data:

- There is at least one jet with $p_T(\text{jet}) > 20\,\text{GeV}$.
- There are no reconstructed leptons meeting the isolation requirement in the final state.
- A set of dedicated jet triggers are used to select the events, as illustrated in Tables 10.5 and 10.6. The data is split into bins based on the η value of the leading jet, and the average p_T of the two leading jets (or the leading jet p_T in the case of a single jet event). The single central jet triggers are used for $\eta < 3.2$ and the single forward jet triggers are used for $\eta > 3.2$. The p_T binning of the triggers correspond to a 99% selection efficiency. As all but two of these triggers are prescaled, each seed event gains a weight corresponding to the trigger prescale of the bin it falls into (Fig. 10.7).
- The final cut is on the E_T^{miss} significance, defined as MET/ΣE_T. As E_T^{miss} resolution varies with the event scalar sum E_T, a cut on E_T^{miss} would potentially bias the p_T scale of the seed events. Hence a cut on E_T^{miss} significance is used. We chose a cut of $\cancel{E}_T^{\text{sig}} < 2.0$, motivated by Fig. 10.8 which shows the lack of any events potentially within our signal region of $E_T^{\text{miss}} > 150\,\text{GeV}$. Figure 10.7 displays kinematic distributions for seed events after trigger selection and isolated lepton veto. Figure 10.9 has the corresponding distributions after applying the $\cancel{E}_T^{\text{sig}}$ requirement, showing the very high purity of this selection.

After selecting the seed events, they need to be smeared using the jet response function described earlier. A "smeared" event is generated by multiplying each jet four-momentum in the seed event by a random number drawn from the jet smearing function. The seed events are split into three regions based on the smeared value of

Table 10.5 Central jet trigger selection if leading jet has $|\eta| < 3.2$

p_T^{avg} threshold (GeV)	Trigger	$\int L dt$	Avg. prescale
90	EF_j55_a4tchad	$0.44\,pb^{-1}$	45,898
125	EF_j80_a4tchad	$2.3\,pb^{-1}$	8754
160	EF_j110_a4tchad	$9.8\,pb^{-1}$	2067
185	EF_j145_a4tchad	$36\,pb^{-1}$	559
240	EF_j180_a4tchad	$79\,pb^{-1}$	257
300	EF_j220_a4tchad	$261\,pb^{-1}$	77.6
380	EF_j280_a4tchad	$1.2\,fb^{-1}$	17.4
480	EF_j360_a4tchad	$20\,fb^{-1}$	1.00

Events where $p_T^{avg} < 90\,\text{GeV}$ have too large a prescale weight and have been left out

Table 10.6 Forward jet trigger selection if leading jet has $|\eta| > 3.2$

p_T^{avg} threshold (GeV)	Trigger	$\int L dt$	Avg. prescale
60	EF_fj45_a4tchad_L2FS	$475\,nb^{-1}$	42,722
90	EF_fj55_a4tchad_L2FS	$511\,nb^{-1}$	39,727
125	EF_fj80_a4tchad	$84\,pb^{-1}$	243
160	EF_fj110_a4tchad	$1.5\,fb^{-1}$	13.3
185	EF_fj145_a4tchad	$12\,fb^{-1}$	1.74
240	EF_fj180_a4tchad	$20\,fb^{-1}$	1.00

Events where $p_T^{avg} < 60\,\text{GeV}$ have too large a prescale weight and have been left out

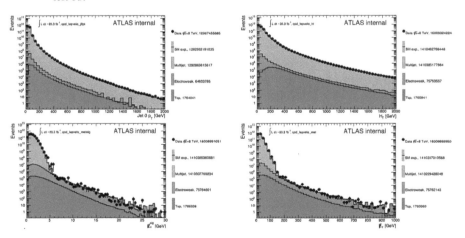

Fig. 10.7 Kinematic distributions for seed events after trigger selection and isolated lepton veto. Note the smooth p_T distributions after applying the prescale weights, and how multijet-enriched the low $\slashed{E}_T$ regions are

$\Delta\phi_{\min}(\slashed{E}_T, j)$. Because most multijet events have $\Delta\phi_{\min}(\slashed{E}_T, j) \leq 1.0$ and all of these events are cut out of the analysis early on, there is no need to generate a large number

10.3 Data-Driven Background Processes

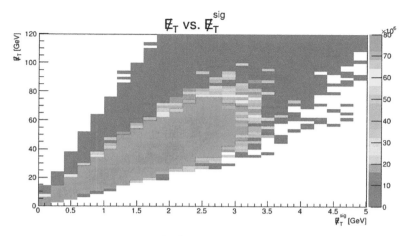

Fig. 10.8 Scatter plot of E_T^{miss} and $E_T^{\text{miss, sig}}$ for seed events after trigger selection and isolated lepton veto

of these smeared events. Each seed event in this region is only smeared ten times, adding an additional weight of 0.1, which provides a suitable number of pseudo-data events. The next region is $1.0 < \Delta\phi_{\text{min}}(\not{E}_T, j) < 1.5$, which is significantly smaller than the other two. It ends up being very unlikely for a seed event to make it to this region, so each event is smeared 1000 times and given a weight of 0.001. The final region is $1.5 \leq \Delta\phi_{\text{min}}(\not{E}_T, j) \leq \pi$, which is the main contribution to multijet events in the signal region. This region is largely populated by 2-jet seed events where 1 jet fails the jet selection criteria after smearing. This is very unlikely to occur, but sufficient statistics is important to the analysis, so this region is smeared 10,000 times and given a weight of 0.0001.

10.3.1.2 Multijet Control Regions

There are three variables which provide excellent separation between multijet and other backgrounds or signal. As shown in Fig. 10.10, the first of these is $\Delta\phi_{\text{min}}(\not{E}_T, j)$ which is the minimum $\Delta\phi$ between the jets and MET. Roughly 90% of multijet events have $\Delta\phi_{\text{min}}(\not{E}_T, j) < 0.7$, and roughly 60% of the events in this region are from the multijet background. The other two variables are E_T^{miss} and the p_T of the leading jet. Multijet events dominate in the low E_T^{miss}, high p_T regions. Cuts on these three variables provide an excellent multijet control region which is orthogonal to the signal region. The final control region chosen was $100 < E_T^{\text{miss}} < 120\,\text{GeV}$ and leading jet $p_T > 100\,\text{GeV}$. This provides a multijet enriched region which is not too biased against the different types of multijet events.

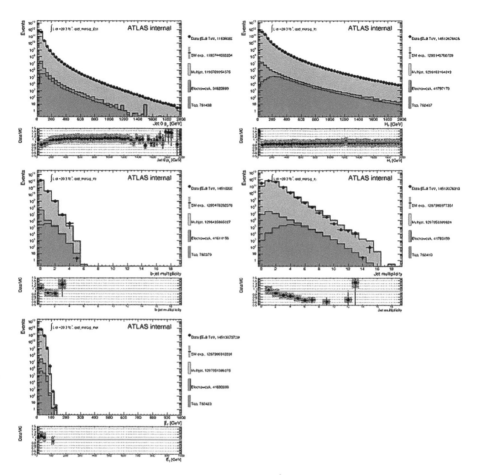

Fig. 10.9 Kinematic distributions for seed events after E_T^{sig} cut. Note that every region is multijet-enriched by many orders of magnitude, even in the high jet multiplicity bins. The $\not{E}_T$ of these events is also under 150 GeV

10.3.1.3 Reweighting

After ϕ and p_T smearing of the jets, there is still slight mismodeling of $\Delta\phi_{\min}(\not{E}_T, j)$ and $p_T(j^1)$, which are important kinematic variables in this analysis. Correcting for this systematic mismodeling is difficult, because there are many different types of multijet events. Mismodeling in the b-jet response function does not necessarily translate to the same mismodeling in the light jet response function, because b-jets contain real E_T^{miss} whereas light jets are typically only fake E_T^{miss}. Additionally, events with a single jet in the final state, where one smeared jet has failed the selection criteria, typically correspond to the far tails of the response function. High p_T dijet events, on the other hand, can easily produce large amounts of E_T^{miss} by

10.3 Data-Driven Background Processes

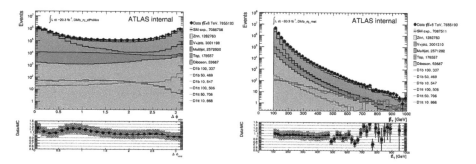

Fig. 10.10 Variables with excellent separation between the multijet background at preselection level minus the $\Delta\phi_{\min}(\slashed{E}_T, j)$ requirement to maintain high efficiency for multijet production. These are $\Delta\phi_{\min}(\slashed{E}_T, j)$, the E_T^{miss}

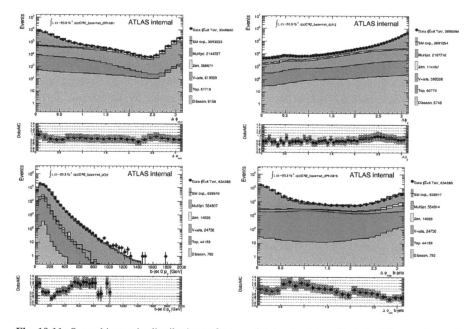

Fig. 10.11 Some kinematic distributions after reweighting by jet multiplicity. The angular and b-jet p_T mis-modeling goes away within errors

small smearing corrections to one of the jets. Taking all of this into account, the multijet sample is reweighted 2-dimensionally with respect to jet multiplicity n_j and b-jet multiplicity n_b. The results of reweighting in the multijet control region can be seen in Fig. 10.11.

For Figs. 10.10 and 10.11, the other non-QCD background processes shown are from MC only, including $Z \rightarrow \nu\bar{\nu}$ which has a very small contribution in the multijet control regions and the MC samples give sufficiently good description in this case.

10.3.2 Estimation of $Z(\to \nu\bar{\nu})$ + jets Background

The production of a vector boson (W or Z) in association with jets constitutes the dominant background in this analysis. The main contribution comes from the irreducible component of $Z+jets$ production in which the Z boson decays into a pair of neutrinos, generating large E_T^{miss}. An accurate description of the normalization and kinematics of this background is crucial to achieve better sensitivity for WIMP pair production in this analysis.

The MC predictions a priori suffer from large theoretical and experimental uncertainties, which affect the absolute normalization and the shape of the predicted distributions. Hence, two complementary data-driven methods are devised to determine the $Z \to \nu\nu$ background. At low E_T^{miss} below 200 GeV, $Z \to \nu\bar{\nu}$ background is estimated from $Z \to \mu\mu$ distribution; at high E_T^{miss} above 200 GeV, we use high p_T γ production for the estimation as it has higher statistics.

10.3.2.1 Data-Driven $Z \to \nu\nu$ Background from $Z \to \mu\mu$

The p_T spectrum of produced Z bosons and the kinematic distributions of jets are the same whether the Z boson decays into charged leptons or neutrinos. Thus the $Z(\to \mu^+\mu^-)$+jets data sample provides the best modeling of the $Z(\to \nu\bar{\nu})$+jets background. To obtain the estimate for $Z \to \nu\nu$ distribution, we reweight $Z \to \mu\mu$ events in data with a transfer function deduced from MC to account for the difference in branching ratio between the two Z decay modes, lepton reconstruction efficiency, trigger and selection efficiency, etc. The data-driven method is carried out in the following steps:

Select $Z \to \mu\mu$ Control Region We define a dedicated control region to select $Z \to \mu\mu$ events. The selection is listed in Table 10.7: "2Muon Preselection" selects the $Z \to \mu\mu$ events; a few additional cuts are used to check the purity of events in

Table 10.7 Selection of the $Z \to \mu\mu$ control region

2Muon preselection	
Muon trigger	
Two tight muons	
Electron veto	
$p_T(\mu_0)$	> 25 GeV
$p_T(\mu_1)$	> 25 GeV
Two muons of opposite charge	
$110 > m_{\ell\ell} > 70$ GeV	
Additional selection	
E_T^{miss} ll	> 100 GeV
Jet multiplicity	$n_{\text{jet}} > 0$
p_T(leading jet)	> 100 GeV
$b$jet multiplicity	$n_b > 0$

10.3 Data-Driven Background Processes

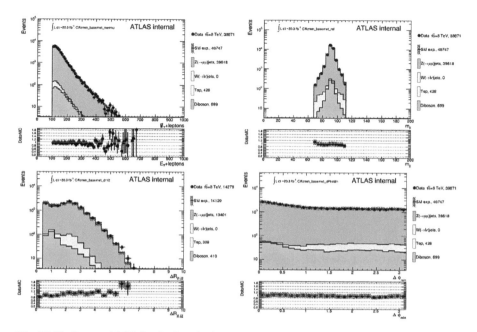

Fig. 10.12 Data and MC distributions in the $Z \to \mu\mu$ control region before b-tag requirement. *Top left* plot is E_T^{miss} ll distributions; *bottom right* plot has E_T^{miss} corrected for E_T^{miss} ll

regions closer to our signal region. For the $Z \to \mu\mu$ distribution, in order to mimic the two neutrino final state, we create a new variable, E_T^{miss} ll, which is the vector sum of the transverse momentum of the two muons and the standard E_T^{miss} vector.

The data and MC distributions in the $Z \to \mu\mu$ CR are shown in Figs. 10.12 and 10.13. The selection has purity of $Z \to \mu\mu$ events about 95 % before b-tagging, and 80 % after requiring one or more b-tagged jets. Good agreement is achieved between data and backgrounds.

Calculate the Transfer Function from $Z \to \mu\mu$ to $Z \to \nu\nu$ Using MC

While the transfer function derived from 2μ data (with respective backgrounds contamination subtracted from MC) performs reasonably well, as used in the DM+heavy-flavor analysis [4], its accuracy is limited by three factors, purity of the samples especially at high E_T^{miss} when the muon selection efficiency decreases, low statistics in data at high E_T^{miss}, and systematic effects as each of the MC samples used in the background subtraction introduces additional systematics to the calculation. In light of the drawbacks, here we adopt an alternative method by deducing the transfer function directly from $Z \to \mu\mu$ and $Z \to \nu\bar{\nu}$ MC samples, which ensures purity of the samples and adequate statistics even at high E_T^{miss}. Additionally, the systematic effects of the MC samples, assuming it is independent of Z decay mode,

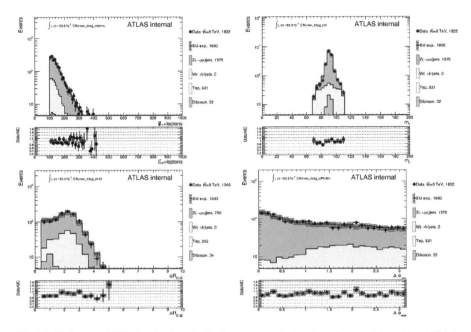

Fig. 10.13 Data and MC distributions in the $Z \to \mu\mu$ control region after requiring at least $1b$ jet. *Top left* plot is E_T^{miss} ll distributions; *bottom right* plot has E_T^{miss} corrected for E_T^{miss} ll

would cancel out in the calculation of the transfer function, when E_T^{miss} distribution from $Z \to \nu\bar{\nu}$ is divided by E_T^{miss} ll from $Z \to \mu\mu$.

The $Z \to \mu\mu$ MC samples are passed through the "2muon preselection" listed in Table 10.7, and the E_T^{miss} is corrected for the muon p_T. The E_T^{miss} trigger in Sect. 8.2.1 is applied to both $Z \to \nu\bar{\nu}$ and $Z \to \mu\mu$ samples, the effect of which would essentially cancel out in the division when calculating the transfer function. The E_T^{miss} trigger is applied in the beginning of the selection stage before additional kinematic cuts, and the calculation of the transfer function is illustrated in Eq. (10.1). The muon trigger is already applied in the selection of $Z \to \mu\mu$ events for E_T^{miss} ll.

$$\text{TF}_{Z \to \mu\mu}^{Z \to \nu\nu} = \frac{(\not{E}_T^{Z \to \nu\nu \text{MC}} * \not{E}_T \text{Trigger}) * \not{E}_T \text{TrigCorr}_{\text{MC}}^{\text{Data}}}{\not{E}_T^{\ell\ell, Z \to \mu\mu \text{MC}} * \not{E}_T \text{Trigger}^{\text{Data}}} \quad (10.1)$$

In Eq. (10.1), the first part of the numerator inside the parenthesis implies that the E_T^{miss} from $Z \to \nu\bar{\nu}$ MC is passed through the E_T^{miss} trigger at the beginning of the selection (prior to additional kinematic cuts). The second part of the numerator is to correct for the small difference in trigger turn on between data and MC as noted in Sect. 8.2.1, and is only applied to E_T^{miss} below 180 GeV as above that the E_T^{miss} trigger reaches full efficiency for both data and MC. Here $\not{E}_T \text{TrigCorr}_{\text{MC}}^{\text{Data}} = \frac{\not{E}_T \text{Trigger}^{\text{Data}}}{\not{E}_T \text{Trigger}^{\text{MC}}}$, where $\not{E}_T \text{Trigger}^{\text{Data}}$ is the trigger curve for data ($W \to \ell\nu$ selection, in black) in

10.3 Data-Driven Background Processes

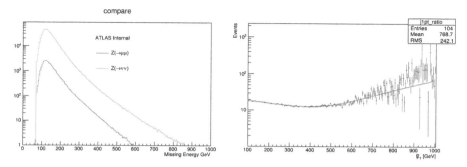

Fig. 10.14 E_T^{miss} ll in $Z \to \mu\mu$ (*blue*) and E_T^{miss} in $Z \to \nu\nu$ (*green*) distributions (*left*) and transfer function of the latter divided by the former (*right*)

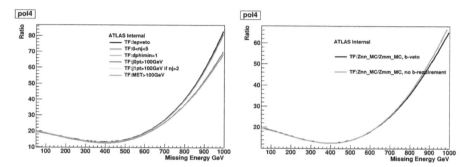

Fig. 10.15 Transfer function varies little across different selection stages (*left*) and b-jet requirements (*right*)

part(a) of Fig. 8.2, and $\slashed{E}_T\text{Trigger}^{\text{MC}}$ is the red curve in the same plot from W+jets MC ($W \to \ell\nu$ selection). The same $\slashed{E}_T\text{Trigger}^{\text{Data}}$ is applied in the denominator of Eq. (10.1) to simulate the E_T^{miss} trigger in $Z \to \mu\mu$ E_T^{miss} ll distribution.

Figure 10.14 shows the $Z \to \mu\mu$ E_T^{miss} ll and $Z \to \nu\bar{\nu}$ E_T^{miss} distributions on the left, and the transfer function calculated as the latter divided by the former on the right. The transfer function is fit by a polynomial to the power of 4 in the region of E_T^{miss} between 50 and 1000 GeV, though in the reweighting process only E_T^{miss} ll below 200 GeV is used.

The transfer function remains consistent across different selection stages for the relevant E_T^{miss} region below 200 GeV, as shown in Fig. 10.15. The variations from the transfer function are taken into account as systematics detailed in Sect. 11.3. To ensure orthogonality to the signal region, the transfer function used for reweighting the $Z \to \mu\mu$ distributions is derived from the region with an anti-btag (b-tagging working point: 60%), and the selection stage after the cut on leading jet p_T is used.

A closure test is performed by reweighting various kinematic distributions, including but not limited to E_T^{miss}, in the $Z \to \mu\mu$ MC sample by the transfer

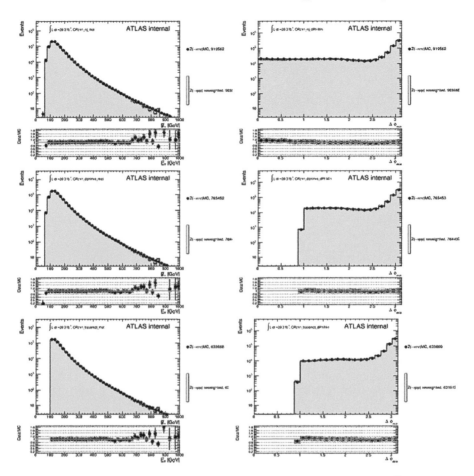

Fig. 10.16 Comparison of reweighted $Z \to \mu\mu$ distributions (*yellow*) and $Z \to \nu\bar{\nu}$ distributions (*dot*) at different selection stages: after n_j cut (*upper*); after $\Delta\phi$min cut (*middle*); after E_T^{miss} cut (*lower*). All samples shown here are MC

function derived from E_T^{miss}, and comparing it to the corresponding distributions in the $Z \to \nu\bar{\nu}$ MC sample. Good agreement is reached as shown in Fig. 10.16.

Data-Driven Estimation of $Z \to \nu\nu$ from $Z \to \mu\mu$ Reweighted by the Transfer Functions

The transfer function is used to reweight the $Z \to \mu\mu$ sample, which is selected from muon-stream data by the selection requirements in Table 10.7. As the purity of the data sample is limited and decreases after *b*-tag requirements, the rest of the background processes (top, diboson and $W(\to \ell\nu)$ + jets; from MC) that passed

10.3 Data-Driven Background Processes

through the same $Z \to \mu\mu$ selection are subtracted from data for a pure sample of $Z \to \mu\mu$ events. The reweighting process is illustrated in Eq. (10.2).

$$Z_{\to\nu\nu}^{\text{data-driven}} = Z_{\to\mu\mu}^{\text{Data-MC}} * \text{TF}_{Z\to\mu\mu}^{Z\to\nu\nu} * \displaystyle{\not}E_T \text{Trigger}^{\text{Data}} \quad (10.2)$$

Here $\text{TF}_{Z\to\mu\mu}^{Z\to\nu\nu}$ is calculated in Eq. (10.1) as a function of E_T^{miss}, and reweighting is performed on an event-by-event basis. As the E_T^{miss} trigger will effectively cancel out in the derivation of the transfer function, $\displaystyle{\not}E_T \text{Trigger}^{\text{Data}}$, which again is the trigger curve for data ($W \to \ell\nu$ selection, in black) in part(a) of Fig. 8.2 is applied to simulate the E_T^{miss} trigger for a more accurate data-driven estimation of $Z \to \nu\bar{\nu}$ distributions.

The same transfer function, derived after the cut on leading jet p_T, is used for reweighting throughout the selection stages. Good agreement is achieved at different selection stages, with potential variation of the transfer function taken into account as a source of systematic uncertainty detailed in Sect. 11.3.

To check the performance of $Z \to \nu\nu$ background modeling after reweighting, we create specific control regions using selections very similar to the final signal selection, but requiring either no b-jet or exactly one b-jet in the final state, so they are orthogonal to the signal region which requires at least two b-jets. To avoid signal contamination, the $1b$-tag CR is blinded by requiring E_T^{miss} below 200 GeV. The selections used for these two CRs are listed in Table 10.8.

We add the obtained $Z \to \nu\nu$ data-driven distributions to the rest of the backgrounds, and compare the overall distributions with data in the aforementioned control region. The agreement is good across all kinematic variables with a slight overestimation of data: we derive a flat scale factor of 0.9, which is applied to the $Z \to \nu\bar{\nu}$ distributions. The deviation from unity in the scale factor, i.e., 10 %, is taken into account in later systematic calculations as uncertainty on the cross-section for the $Z(\to \nu\bar{\nu})$+jets process. Figure 10.17 shows the distributions in the b-jet veto region, and Fig. 10.18 shows the distributions in the $1b$-tag region, in both cases with

Table 10.8 Selection of the $Z \to \nu\bar{\nu}$ control region

E_T^{miss} trigger	
Lepton veto	
No b-jet or exactly one b-jet with b-tagging working point: 60 %	
E_T^{miss}	> 100 GeV
if $n_b = 1$	E_T^{miss} < 200 GeV (blind)
Jet multiplicity	$1 \leq n_j \leq 4$
$\Delta\phi_{\min}(E_T^{\text{miss}}, p_T(j))$	> 1.0
p_T(leading jet)	> 100 GeV
if $n_j = 3$	p_T(sub-leading jet) > 100 GeV
Jet multiplicity	$2 \leq n_j \leq 3$
$\Delta\phi(p_T(j_0), p_T(j_1))$	< 2.0
$90 \text{ GeV} \leq m_{jj} \leq 150 \text{ GeV}$	

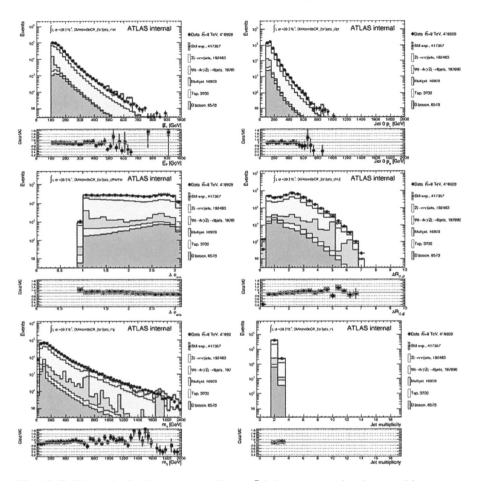

Fig. 10.17 Kinematic distributions in the $Z \to \nu\bar{\nu}$ b-jet veto control region, requiring two or three jets in the final state. A scale factor of 0.9 is applied to the $Z \to \nu\bar{\nu}$ estimation, which is from reweighted $Z \to \mu\mu$ only. Some statistical effect is observed in the multijet distribution due to the limited number of smeared events, which does not affect the overall background description as most multijet events are removed by subsequent selection requirements. An earlier selection stage is shown here for sufficient statistics in different processes

the scale factor applied. As seen in the modeling for most kinematic variables, this reweighting method yields a very good description of the $Z \to \nu\nu$ background, and the good agreement between data and backgrounds is preserved through different selection stages.

10.3 Data-Driven Background Processes

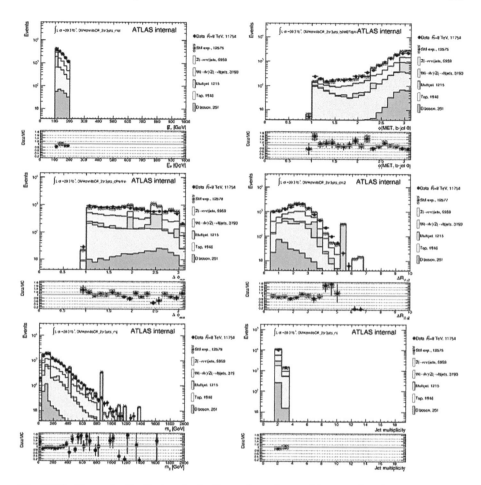

Fig. 10.18 Kinematic distributions in the $Z \to \nu\bar{\nu}$ one b-tagged jet control region, requiring two or three jets in the final state, and blinded by requiring $E_T^{\text{miss}} < 200$ GeV. A scale factor of 0.9 is applied to the $Z \to \nu\bar{\nu}$ estimation, which is from reweighted $Z \to \mu\mu$ only. Some statistical effect is observed in the multijet distribution due to the limited number of smeared events, which does not affect the overall background description as most multijet events are removed by subsequent selection requirements. An earlier selection stage is shown here for sufficient statistics in different processes

Data-Driven $Z(\to \nu\bar{\nu})$ + Jets Background from γ+Jets

The above approach provides a very accurate description of the $Z \to \nu\nu$ background. However there are over nine times more $Z \to \nu\nu$ events than $Z \to \mu\mu$ due to their relative branching ratios and acceptances. This means that the data-driven $Z \to \nu\nu$ estimate described in Sect. 10.3.2.1 will always have about nine times fewer events than the background it is trying to simulate. In the signal region at high E_T^{miss}, where the total number of events is highly suppressed, low statistics in $Z \to \mu\mu$ data leads to non-negligible inaccuracies. The solution chosen in this analysis is to use the production of high energy γ-rays to model Z production, which has the advantages of a larger cross-section and stability. For γ transverse momenta much greater than the mass of the Z boson, the kinematics of γ + jets and Z+jets are very similar [5].

Comparison of $Z \to \nu\bar{\nu}$ and γ Distributions Similar to the E_T^{miss} ll definition for the $Z \to \mu\mu$ sample, we define E_T^{miss} g as the vector sum of the E_T^{miss} vector and the γ-ray p_T. Throughout the analysis, any reference to E_T^{miss} of an event from the γ+jets selection is taken to mean E_T^{miss} g.

For a detailed description of the photon definition and selection, see Chap. 8. The lowest unprescaled photon trigger is the g120_loose trigger, which has a near 100 % efficiency plateau at 125 GeV p_T. To obtain an enriched sample of high p_T photons, we require the following cuts:

- g120_loose fired
- Exactly 1 photon
- Photon $p_T > 125$ GeV
- Lepton veto on electrons and muons
- E_T^{miss} g, the vector sum of the photon p_T and the E_T^{miss}, is larger than 100 GeV.

The purity of this selection is about 85 %. To study the sources of the contamination, we plot the backgrounds that pass the γ selections above. As γ + jets MC sample does not describe the shape of γ + jets data well, in Figs. 10.19 and 10.20, the data is the γ data stream: the same γ + jets data is also stacked with other backgrounds (from MC) and the sum is compared with γ + jets data alone; the deviations from 1 in the ratio on the lower pads reflect the amount of contamination of other backgrounds that pass the γ selection. As Fig. 10.19 shows, most of this contamination comes from multijet background at $\Delta\phi_{\min}(E_T^{\text{miss }\gamma}, p_T(j)) < 2.0$ and low E_T^{miss} g. By placing two additional cuts requiring $\Delta\phi_{\min}(E_T^{\text{miss }\gamma}, p_T(j)) > 2.0$ and E_T^{miss} g > 200 GeV to remove multijet background, the purity of the sample can be brought to 95 % before b-tagging, as is shown in Fig. 10.20. As b-tagging further removes most of the gluon background, where a jet may fake a photon, (see lower left plot of Fig. 10.20) the sample has over 99 % purity in the nominal cutflow after b-tagging, giving us a very clean γ + jets data sample to be used for reweighting in our signal region.

Because of the 125 GeV p_T cut on the photon, the two estimates are mismatched for low E_T^{miss} where Z bosons with $p_T < 125$ GeV contribute significantly. Additionally, kinematics of the two processes are significantly different close to Z mass, and the productions only start resembling each other at energy significantly above the Z mass. As illustrated in Fig. 10.21, for $E_T^{\text{miss}} > 200$ GeV, the γ + jets sample models the Z background very well.

Calculate the Transfer Function from γ to $Z \to \nu\bar{\nu}$ The next step in this method is to fit a transfer function from γ +jets sample to the $Z \to \nu\bar{\nu}$ selection. For similar reasons described in the $Z \to \mu\mu$ method, MC samples are used in the derivation of

10.3 Data-Driven Background Processes

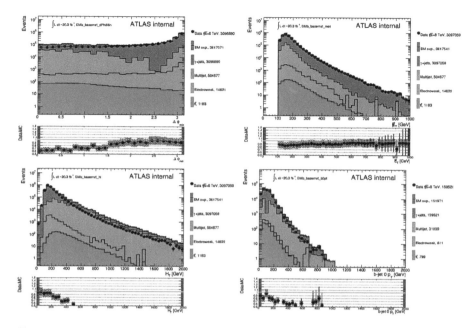

Fig. 10.19 Various kinematic distributions at photon preselection **before** multijet cleaning cuts

the transfer function, for both $Z \to \nu\bar{\nu}$ and γ samples. We use a physics-motivated fit function to fit a $Z \to \nu\bar{\nu}$ to $\gamma + \text{jets}$ transfer function. The $\gamma + \text{jets}$ sample is then divided by this function to obtain the $Z \to \nu\bar{\nu}$ estimate.

The $\gamma + \text{jets}$ MC sample is passed through the same high p_T photon selection as detailed above, and its E_T^{miss} corrected to E_T^{miss} g. A lepton veto is applied to both $\gamma + \text{jets}$ and $Z \to \nu\bar{\nu}$ MC samples, and the E_T^{miss} trigger in Sect. 8.2.1 is applied to both $Z \to \nu\bar{\nu}$ and $\gamma + \text{jets}$ samples, the effect of which would essentially cancel out in the division when calculating the transfer function. To check robustness of the method, a series of kinematic cuts as listed in Table 10.9 are applied consecutively, and a transfer function is derived at each stage: the cuts are consistent with the other CRs, and the SR at preselection level (before b-tag requirement).

Figure 10.22 shows the $Z \to \nu\bar{\nu}$ and $\gamma + \text{jets}$ E_T^{miss} distributions on the left at the selection stage after requiring $\Delta\phi_{\min}(E_T^{\text{miss}}, p_T(j)) > 1.0$ and the transfer function calculated as the latter divided by the former on the right. Similar to Eq. (10.1), the application of the E_T^{miss} trigger in the calculation of the transfer function is illustrated in Eq. (10.3). The E_T^{miss} trigger is applied in the beginning of the selection stage before additional kinematic cuts, and the γ trigger is already applied in the selection of $\gamma + \text{jets}$ events to obtain E_T^{miss} g.

$$\text{TF}^{\gamma}_{Z \to \nu\nu} = \frac{\displaystyle \not{E}_T^{\gamma,\gamma MC} * \not{E}_T \text{Trigger}^{\text{Data}}}{(\not{E}_T^{Z \to \nu\nu MC} * \not{E}_T \text{Trigger}) * \not{E}_T \text{TrigCorr}^{\text{Data}}_{\text{MC}}} \quad (10.3)$$

126   10 Dark Matter + Higgs($\to b\bar{b}$): Background Processes

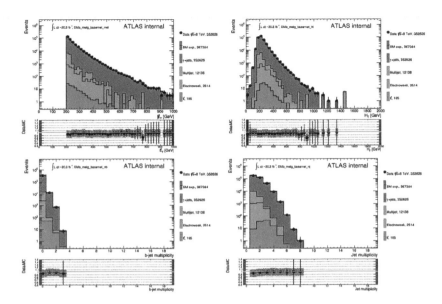

Fig. 10.20 Various kinematic distributions at photon preselection after multijet cleaning cuts have been applied. Note the b-jet multiplicity plot shows that b-tagging requirement will remove most of the gluon and light jet background

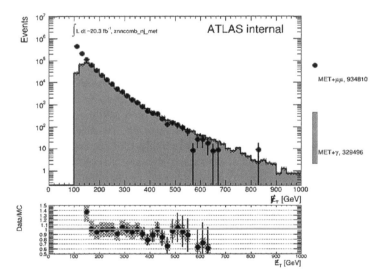

Fig. 10.21 E_T^{miss} shape comparison of the two estimates at preselection

The transfer function is fit with Eq. (10.4) in the region 200–1000 GeV. The function [6] is based on the production mechanism of the two processes, including the expected Z mass suppression relative to the photon distribution for $p_T < M_Z$. Theoretical calculations give $n \sim 1.2$, while our fit yields $n \sim 1.3$ very close to

10.3 Data-Driven Background Processes

Table 10.9 Selections used to calculate the transfer function from γ + jets to $Z \to \nu\bar{\nu}$

E_T^{miss} trigger	
Lepton veto	
E_T^{miss}	> 100 GeV
Jet multiplicity	$1 \leq n_j \leq 4$
$\Delta\phi_{min}(E_T^{miss}, p_T(j))$	> 1.0
p_T(leading jet)	> 100 GeV
if $n_j > 2$	p_T(sub-leading jet) > 100 GeV

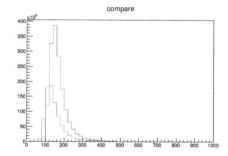
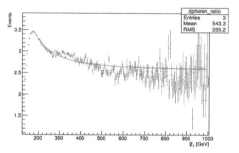

Fig. 10.22 E_T^{miss} g in γ (*blue*) and E_T^{miss} in $Z \to \nu\nu$ (*red*) distributions (*left*) and transfer function of the former divided by the latter (*right*)

predicted value. The $\chi^2/$dof for this fit is 2.62 for $n_{dof} = 158$. The small fluctuations of the shape of the ratio from that of the transfer function are taken into account as systematics detailed in Sect. 11.3.

$$f(x) = R((x^2 + M_Z^2)/x^2)^n \tag{10.4}$$

The transfer function is calculated for different selection stages in Table 10.9: the TF remains consistent across different selection stages, as shown in Fig. 10.23 (left). The variations from the transfer function is taken into account as systematics detailed in Sect. 11.3. To ensure orthogonality to the signal region, the transfer function used for reweighting the γ + jets distributions is derived from the region with an anti-btag (*b*-tagging working point: 60 %), and the selection stage after the cut on $\Delta\phi_{min}(E_T^{miss}, p_T(j))$ is used.

As the coupling to quarks differs between Z and γ, we study the effects of the difference by deriving the transfer functions with different *b*-jet multiplicity requirements, and comparing them by dividing one transfer function against another. As shown in Fig. 10.24, the ratio of the transfer functions is flat at E_T^{miss} above 200 GeV, so we can apply a simple scale factor (scale factor) to account for the difference in coupling to quarks between Z and γ. To derive the scale factor, we first divide the number of events as a function of *b*-jet multiplicity in γ + jets by that of $Z \to \nu\bar{\nu}$, as shown in Fig. 10.25. As the transfer function is derived from a *b*-veto region, we then divide the bin content corresponding to the specific *b*-jet

10 Dark Matter + Higgs($\to b\bar{b}$): Background Processes

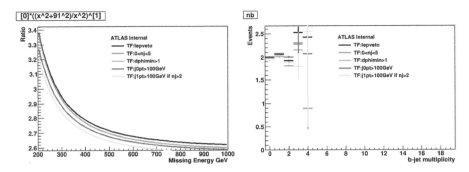

Fig. 10.23 Little variation across different selection stages for transfer function (*left*) and ratio of b-jet multiplicity (*right*)

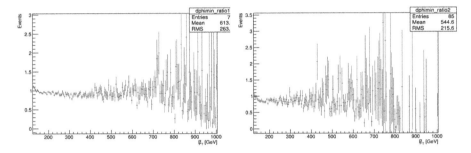

Fig. 10.24 Transfer function derived at b-veto stage divided by that from exactly one b-jet stage (*left*) and exactly two b-jets stage (*right*)

multiplicity, e.g., one-b-jet, by the bin content in the zero-b-jet bin, and that gives us the scale factor. As seen in the right plot in Fig. 10.23, the ratio in b-jet multiplicity is also relatively consistent across different selection stages: most of the events in the analysis have no more than 2 b-jets, leading to larger statistical fluctuations in the $n_b = 3$ and $n_b = 4$ bins in the plot.

The reweighting process is illustrated in Eq. (10.5).

$$Z^{\text{data-driven}}_{\to \nu\nu, n_b=i} = \frac{\gamma^{\text{Data}}}{\text{TF}^{\gamma}_{Z \to \nu\nu} * \text{scalefactor}_{n_b=i}} * \cancel{E}_T \text{Trigger}^{\text{Data}} \qquad (10.5)$$

Here $\text{scalefactor}_{n_b=i} = \frac{\text{Ratio}^{n_b=i}_{\text{evt}}}{\text{Ratio}^{n_b=0}_{\text{evt}}}$, and $\text{Ratio}^{n_b=i}_{\text{evt}}$ is the ratio of the number of γ +jets events divided by that of $Z \to \nu\bar{\nu}$ as a function of b-jet multiplicity, i.e., the bin content for $n_b = i$ in Fig. 10.25. The effect of b-tagging efficiency is taken into account as part of the systematic uncertainties detailed in Sect. 11.3.

Here $\text{TF}^{\gamma}_{Z \to \nu\nu}$ is calculated in Eq. (10.1) as a function of E^{miss}_T, and reweighting is performed on an event-by-event basis. As the E^{miss}_T trigger will effectively cancel out in the derivation of the transfer function, $\cancel{E}_T \text{Trigger}^{\text{Data}}$, which again is the

10.3 Data-Driven Background Processes

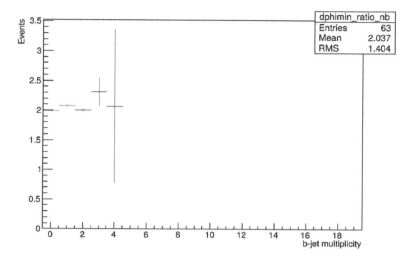

Fig. 10.25 Ratio of the number of events as a function of b-jet multiplicity between $\gamma +$ jets and $Z \to \nu\bar{\nu}$

trigger curve for data ($W \to \ell\nu$ selection, in black) in part(a) of Fig. 8.2 is applied to simulate the E_T^{miss} trigger for a more accurate data-driven estimation of $Z \to \nu\bar{\nu}$ distributions.

A MC closure test is performed by reweighting various kinematic distributions, including but not limited to E_T^{miss}, in the $\gamma +$ jets MC sample by the transfer function derived from E_T^{miss}. Given the poor matching of $\gamma +$ jets events compared to $Z \to \mu\mu$ events for $E_T^{miss} < 180\,\text{GeV}$, low E_T^{miss} region below 200 GeV is reweighted using $Z \to \mu\mu$ MC samples as shown in the previous closure test in Fig. 10.16. For $E_T^{miss} > 200\,\text{GeV}$, $\gamma +$ jets MC sample is used for the reweighting. The combined reweighted results are compared to the corresponding distributions in the $Z \to \nu\bar{\nu}$ MC sample. Good agreement is reached as shown in Fig. 10.26.

Apply the Transfer Function from γ to $Z \to \nu\bar{\nu}$ The transfer function from γ to $Z \to \nu\nu$ as shown in Fig. 10.22 is then applied to the $\gamma +$ jets data sample to get the final estimate of $Z \to \nu\bar{\nu}$ background.

A direct comparison of $Z \to \nu\bar{\nu}$ estimate from $\gamma +$ jets and that from $Z \to \mu\mu$, in the overlap region of E_T^{miss} between 200 and 300 GeV is shown in Fig. 10.27. Good agreement is seen between the two estimates, justifying the use of the $\gamma +$ jets events to estimate $Z \to \nu\nu$ background.

Combined $Z \to \nu\nu$ Background Determination

Given the poor matching of $\gamma +$ jets events compared to $Z \to \mu\mu$ events for $E_T^{miss} < 180\,\text{GeV}$, the low E_T^{miss} region below 200 GeV is estimated using only the original $Z \to \mu\mu$ selection from 2μ data (with other backgrounds subtracted from MC). For $E_T^{miss} > 200\,\text{GeV}$ both methods provide accurate descriptions of the $Z \to \nu\nu$

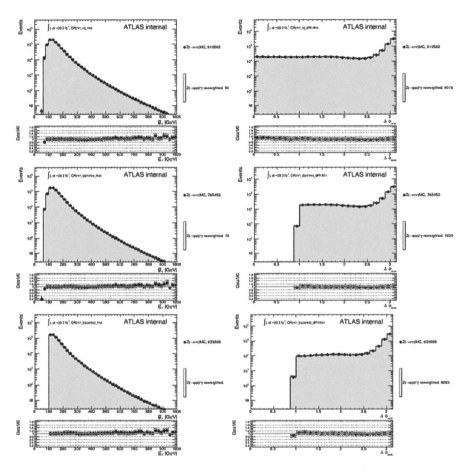

Fig. 10.26 Comparison of reweighted γ + jets distributions (*yellow*) for $E_T^{miss} > 200\,\text{GeV}$ (plus $Z \to \mu\mu$ distributions for $E_T^{miss} < 200\,\text{GeV}$) and $Z \to \nu\bar{\nu}$ distributions (*dot*) at different selection stages: after n_j cut (*upper*); after $\Delta\phi$min cut (*middle*); after E_T^{miss} cut (*lower*). All samples shown here are from MC simulation

background. As the $Z \to \nu\bar{\nu}$ estimation from γ + jets provides over 30 times more statistics covering the high E_T^{miss} regions, it is used for E_T^{miss} over 200 GeV.

We add the combined $Z \to \nu\nu$ data-driven distribution from both methods to the rest of the backgrounds, and compare the overall distributions with data in the aforementioned $Z \to \nu\bar{\nu}$ control region. Figure 10.28 shows the distributions in the b-veto region after the cut on the leading jet p_T. The agreement is good across all kinematic variables with a slight overestimation of data similar to what was observed earlier reweighting using only $Z \to \mu\mu$: as this region is dominated by $Z \to \nu\bar{\nu}$, by fixing the normalization on the non-$Z \to \nu\bar{\nu}$ background, and matching the integrated event count of $Z \to \nu\bar{\nu}$ events to that of data for different E_T^{miss} regions of concern, as detailed in Table 10.10, we derive a flat scale factor of 0.9, which is

10.4 0-Lepton Validation Region

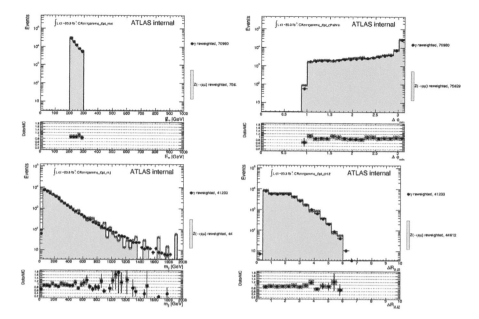

Fig. 10.27 Comparison of $Z \to \nu\bar{\nu}$ estimation from γ + jets (data, *dot*) and $Z \to \mu\mu$ (data with other background contamination subtracted from MC, *yellow*) in the overlap region of E_T^{miss} between 200 and 300 GeV

applied to the $Z \to \nu\bar{\nu}$ distributions across the entire E_T^{miss} range. The scale factor is taken into account in later systematic calculations. Detailed description of the systematics evaluation for $Z \to \nu\bar{\nu}$ estimation is in Sect. 11.3.

Figure 10.29 shows selected kinematic distributions in $Z \to \nu\bar{\nu}$ b-veto control region after further kinematic selections with the scale factor applied. Good agreement is reached across different kinematic variables and selection stages. As the $Z \to \nu\bar{\nu}$ 1b-tag control region is blinded with $E_T^{miss} < 200$ GeV, its $Z \to \nu\bar{\nu}$ contribution is only from reweighted $Z \to \mu\mu$, and is shown in Fig. 10.18 earlier with good agreement across selection stages and kinematic variables.

10.4 0-Lepton Validation Region

After obtaining and validating the individual background processes in Sects. 10.2 and 10.3, we apply them in the 0-lepton validation region to examine the overall description of data. The scale factor of 0.92 from the 1-lepton W+jets control region (Sect. 10.2.1) is applied here to $W(\to \ell\nu)$ & $Z(\to \ell\ell)$ + jets events, and the scale factor of 0.9 from the zero-b-jets $Z(\to \nu\bar{\nu})$ + jets control region (Sect. 10.3.2) is applied here to $Z \to \nu\bar{\nu}$ estimation. The effects of both scale factors are taken into account in subsequent systematics estimation.

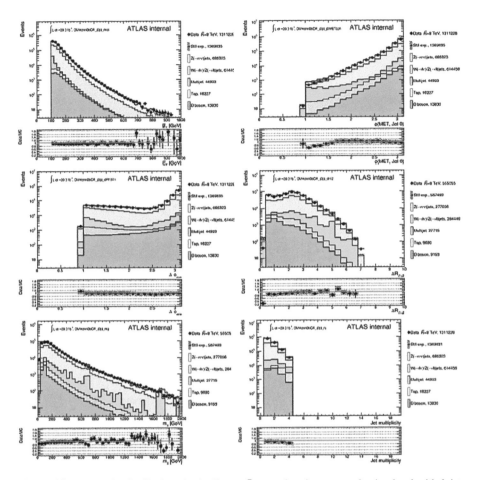

Fig. 10.28 Kinematic distributions in the $Z \to \nu\bar{\nu}$ control region at preselection level with b-jet veto, after cut on leading jet p_T. $Z \to \nu\bar{\nu}$ estimation is from both reweighted $Z \to \mu\mu$ and $\gamma +$ jets combined. A scale factor of 0.9 is derived (but not applied) from the data/bkgd ratio

The 0-lepton validation region has near identical selections to the signal region (Sect. 9.2), with the cut on the invariant mass of the two leading b-jets (m_{bb}) reversed to require $m_{bb} < 60$ GeV or $m_{bb} > 150$ GeV. To preserve statistics, we choose the full E_T^{miss} range above 100 GeV, and ease the cut on $\Delta R(p_T(j0), p_T(j1))$ to below 2.0 instead of 1.5 as is in the signal selection. The selections for 0-lepton control region are listed as follows (the preselection cuts listed in Sect. 9.1 are applied as well):

- Lepton Veto
- $m_{bb} < 60$ GeV or $m_{bb} > 150$ GeV
- $E_T^{\text{miss}} > 100$ GeV
- $\Delta\phi_{\min}(E_T^{\text{miss}}, p_T(j)) > 1.0$
- highest p_T jet must have $p_T > 100$ GeV
- $\geq 1b$-jet with b-tagging working point: 60 %

10.4 0-Lepton Validation Region

Table 10.10 Integrated event count in different E_T^{miss} regions for data, non-$Z \to \nu\bar{\nu}$ background and $Z \to \nu\bar{\nu}$

E_T^{miss} (GeV)	$100 < E_T^{miss} < 200$	$150 < E_T^{miss} < 200$	$E_T^{miss} > 200$	$E_T^{miss} > 300$	$E_T^{miss} > 400$
Data	1.17E06	3.17E05	1.38E05	1.93E04	4.00E03
Non-$Z \to \nu\bar{\nu}$ bkgd	6.26E05	1.53E05	5.75E04	7.10E03	1.40E03
$Z \to \nu\bar{\nu}$ (from $Z \to \mu\mu$)	5.98E05	1.84E05	–	–	–
$Z \to \nu\bar{\nu}$ (from γ + jets)	–	–	8.79E04	1.30E04	2.78E03
Scale factor for $Z \to \nu\bar{\nu}$	0.91	0.89	0.92	0.94	0.94

The scale factors are derived by fixing the normalization on other backgrounds and matching $Z \to \nu\bar{\nu}$ to data. A flat scale factor of 0.9 is adopted in the final estimation of $Z \to \nu\bar{\nu}$ background

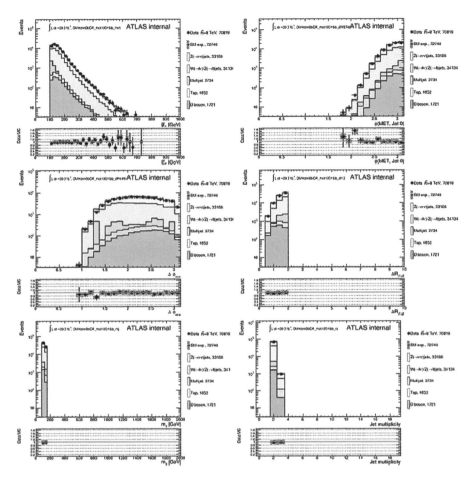

Fig. 10.29 Kinematic distributions in the $Z \to \nu\bar{\nu}$ b-jet veto control region, requiring two or three jets in the final state, and the invariant mass of the two leading jets be between 90 and 150 GeV. A scale factor of 0.9 is applied to the $Z \to \nu\bar{\nu}$ estimation, which is from reweighted $Z \to \mu\mu$ in E_T^{miss} below 200 GeV, and γ + jets for E_T^{miss} above 200 GeV. Systematic uncertainties are given as hatched band and statistical uncertainties as error bars

- highest p_T b-jet must have $p_T > 100$ GeV
- $2 \leq n_j \leq 3$
- sub-leading jet $p_T > 100$ GeV when $n_j = 3$
- sub-leading b-jet $p_T > 60$ GeV when $n_j = 3$ and $n_b > 1$
- $\geq 2b$-jet with b-tagging working point: 60 %
- $\Delta R(p_T(j0), p_T(j1)) < 2.0$

Comparison between data and background is performed for different kinematic distributions including jet p_T, angular distributions, multiplicity, and E_T^{miss}. Figure 10.30 shows the comparison of data and expected background at the selection

10.4 0-Lepton Validation Region

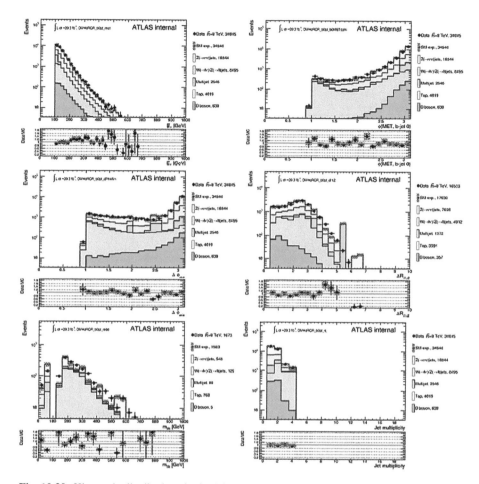

Fig. 10.30 Kinematic distributions in the 0-lepton validation region at preselection level with at least 1-btagged jet, with signal region blinded by reversing the cut on the invariant mass of the two leading b-jets. Systematic uncertainties are given as hatched band and statistical uncertainties as error bars

stage after one b-tag and jet p_T cuts, but prior to the second b-jet requirement and jet multiplicity requirement, for the full range of E_T^{miss} above 100 GeV (excluding the blinded m_{bb} region). Figure 10.31 shows the comparison of data and expected background after imposing the additional jet multiplicity requirement ($2 \leq n_j \leq 3$) with at least one b-jet for the full range of E_T^{miss} above 100 GeV (excluding the blinded m_{bb} region). Figure 10.32 shows the comparison of data and expected background contributions after final selection with at least two b-tagged jets (excluding the blinded signal mass region): to preserve statistics, the plots shown are for the full E_T^{miss} range above 100 GeV, and the cut on $\Delta\phi(p_T(j0), p_T(j1))$ is eased to be ≤ 2.0. Good agreement is achieved for various selections stages.

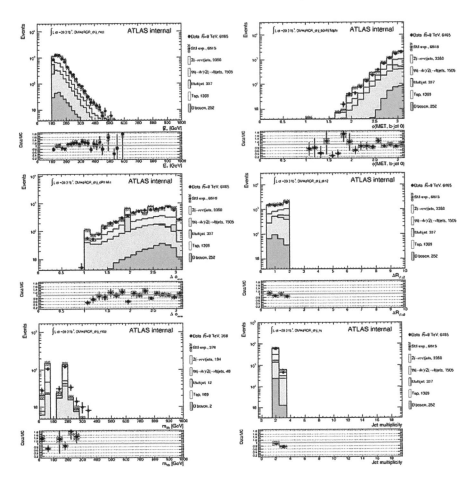

Fig. 10.31 Kinematic distributions in the 0-lepton validation region at preselection level with at least 1-btagged jet, $2 \leq n_j \leq 3$, $\Delta R(j_0, j_1) < 2.0$, with signal region blinded by reversing the cut on the invariant mass of the two leading b-jets. Systematic uncertainties are given as hatched band and statistical uncertainties as error bars

References

1. ATLAS Collaboration, Measurements of normalized differential cross sections for $t\bar{t}$ production in pp collisions at $\sqrt{s} = 7$ TeV using the ATLAS detector. Phys. Rev. **D90**, 072004 (2014)
2. ATLAS Collaboration, Search for the Standard Model Higgs boson produced in association with top quarks and decaying into $b\bar{b}$ in pp collisions at $\sqrt{s} = 8$ TeV with the ATLAS detector. Eur. Phys. J. **C75**(7), 349 (2015)
3. S. Owen, Data-driven estimation of the QCD multijet background to SUSY searches with jets and missing transverse momentum at ATLAS using jet smearing. Technical Report ATL-PHYS-INT-2012-008, CERN, Geneva, Feb 2012

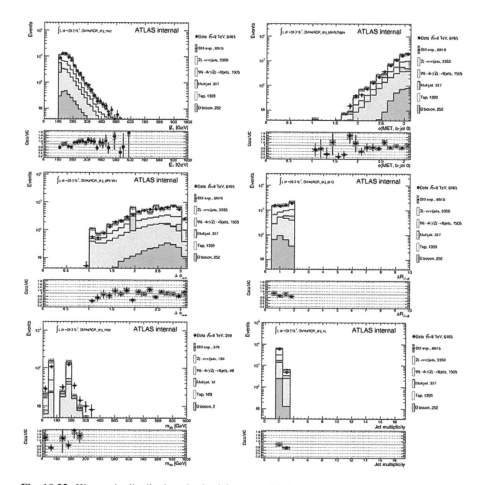

Fig. 10.32 Kinematic distributions in the 0-lepton validation region after full selection (at least two b-tagged jets, $2 \leq n_j \leq 3$, $\Delta R(j_0, j_1) < 2.0$), with signal region blinded by reversing the cut on the invariant mass of the two leading b-jets. Systematic uncertainties are given as hatched band and statistical uncertainties as error bars

4. H. Abreu, G. Artoni, Y. Cheng, A. DiMattia, E. Gozani, Y.-K. Kim, M. Kruskal, P. Priscilla, S. Pataraia, B. Penning, E. Rosenfeld, Y. Rozen, S. Schramm, G. Sciolla, M. Shochet, A. Venturini, Search for Dark Matter in association with b quarks and missing transverse energy in proton-proton collisions at $\sqrt{s} = 8$ TeV with the ATLAS detector. Technical Report ATL-COM-PHYS-2014-217, CERN, Geneva, March 2014
5. S. Ask, M.A. Parker, T. Sandoval, M.E. Shea, W.J. Stirling, Using γ+jets production to calibrate the standard model $Z(\nu\nu)$+jets background to new physics processes at the LHC. J. High Energy Phys. **10**, 058 (2011)
6. T. Sandoval, M.A. Parker, S. Ask, Estimation of the $Z \to \nu\nu$ background to new physics searches in ATLAS. Ph.D. thesis, Cambridge University, May 2012. Presented 22 June 2012

Chapter 11
Dark Matter + Higgs($\to b\bar{b}$): Systematic Uncertainties

This chapter discusses the systematic uncertainties associated with the Monte Carlo (MC) simulated samples used in this analysis. The simulated signal samples are discussed in Sect. 7.3, and the background samples from simulation are given in Sect. 8.1.2. For definitions and reconstruction method of the physics objects used in this analysis, see Chap. 8. The systematic uncertainty figures are used in the final statistical interpretation and limit setting in Chap. 12.

This chapter is organized as follows: Sect. 11.1 describes the main sources of systematic uncertainties and how they are calculated; Sect. 11.2 gives the theoretical uncertainties associated with the signal samples; Sect. 11.3 describes the theoretical and experimental uncertainties associated with the data-driven estimation of the $Z(\to \nu\bar{\nu})$ + jets process; and finally the systematic uncertainty tables and summary are given in Sect. 11.4 for the signal processes, individual background processes, and combined.

11.1 Sources of Systematic Uncertainties

The uncertainty of a measurement is given by a range of values which are likely to enclose the true value. There are two classes of uncertainties which may impact the final result. The first sort of uncertainties are statistical uncertainties which are a result of stochastic fluctuations. They arise from the fact that a measurement is based on a finite set of observations and can be handled using the well-developed mathematical theory of statistics. The second class of uncertainties is the systematic uncertainties. They arise from uncertainties associated with the nature of the measurement apparatus, methods used for the readout or the particle reconstruction, model-dependent uncertainties such as the choice and parameterization of MC simulations, and further assumptions made. The evaluation of systematic uncertainties is generally not trivial and will be discussed in the following.

The uncertainties on various experimental quantities used in the analysis may have a sizable effect on the result and have to be investigated carefully.

- *Rate Uncertainties*: Uncertainties related to efficiencies and overall normalization of the contributing physical processes. This class of uncertainties is not changing the shape of the distribution used for limit settings but the ratio of the various contributions to the entire data set.
- *Shape Uncertainties*: If the variation of a source of systematic uncertainty changes the value of a given quantity differently for different events. These uncertainties are referred to as shape or shape-changing uncertainties.

The rate uncertainties are determined by propagating the systematic variation through the selection requirements as described and calculating the relative uncertainty. For the shape uncertainties the variations are propagated through the selection requirements. By comparing the non-modified ("nominal") and modified shapes, the fractional uncertainties are derived. The integral of the fractional uncertainty gives the overall relative uncertainty. Currently we consider all systematics as rate uncertainties only because our limit setting does not depend on the shape of distributions.

The following sources of systematic uncertainties are considered.

Heavy-Flavor Tagging The b-tagging efficiencies have been determined from comparisons between simulation and heavy-flavor-enriched data samples [1]. The scale factors from the method are used to calculate the systematic uncertainties here. The scale factors are determined independently for b-jets, c-jets and light jets, and their uncertainties are uncorrelated. Therefore the b-tagging uncertainty is calculated separately for each kind of jet, by varying the b-jet(c-jet or miss-tag rate) efficiency SF in all p_T, η bins up (down) by 1σ, and varying the corresponding inefficiency SF in all p_T, η bins down (up) by 1σ at the same time. The final systematic uncertainty due to the b-tagging is obtained by summing these three uncertainties in quadrature. The systematic uncertainties associated with heavy-flavor tagging are sample dependent and range from $\sim$10 % to 15 %.

The uncertainty on the c-jet and light jet scale factors are also calculated propagating the estimated uncertainties on the scale factors to the event weights up or down 1σ. The c-jet and light jet-related uncertainties are small at 3 % or less.

Jet Energy Scale and Resolution The jet energy is scaled up and down (in a fully correlated way) by the 1σ uncertainty on the jet energy scale obtained using the tool [2]. The package JetUncertainties-00-08-07 has been used. This tool returns a relative uncertainty on the jet energy scale which is the sum in quadrature of five components: (1) an uncertainty dependent on the transverse momentum and pseudorapidity of the jet, (2) an uncertainty dependent on the ΔR of the closest jet, (3) an uncertainty dependent on the average quark-gluon composition of the sample, (4) an uncertainty on the pile-up subtraction procedure, and (5) an additional

11.1 Sources of Systematic Uncertainties

uncertainty specific for b-jets. The jet four-vectors are scaled according to the uncertainty and the E_T^{miss} recomputed by using the MissingETUtility package. The uncertainty associated with jet energy resolution is calculated in a similar way, but only in one direction as the best energy resolution is the nominal one used for the analysis, and the uncertainty is obtained by increasing it by 1 σ and compare the effects. The systematic uncertainty associated with the jet energy scale and jet energy resolution range between $\sim$5 % and 15 %.

In top events, mainly consisting of $t\bar{t}$, there is a larger tail at higher p_T compared with other background processes, leading to considerable statistical fluctuations in the jet energy scale uncertainty calculation. A similar effect is observed in the $Z \to \nu\bar{\nu}$ estimation at E_T^{miss} greater than 150 GeV, where top background (from MC) is subtracted from 2μ data for reweighting. The jet energy scale uncertainties for top and $Z \to \nu\bar{\nu}$ with $E_T^{miss} \geq 150$ GeV are hence symmetrized to mitigate the statistical effects.

E_T^{miss} Calculation The impact of the cluster energy uncertainties on the reconstructed E_T^{miss} is estimated using the MissingETUtility tool. The tool is used to propagate the object systematics and to retrieve the systematics on the global E_T^{miss} object. The relevant systematic variations are performed by varying the soft term energy scale and resolution. The typical size of the uncertainty is about $\approx$1 %. The resolution systematic is one sided and symmetrized to obtain the up/down variations.

Pile-Up Simulated events are generated and reweighted with templates matching the luminosity, average number of collisions per bunch crossing, and pile-up conditions. The default pile-up reweighting applies a scale factor of 1.09 to simulated events. The scale factor is varied to 1.05 and 1.13 to obtain the down/up variations of pile-up systematics. The systematic uncertainties associated with pile-up reweighting are small at below 1 %.

Jet Vertex Fraction The jet vertex fraction (JVF) allows for the identification and selection of jets originating in the hard-scatter interaction through the use of tracking and vertexing information. By combining tracks and their primary vertices with calorimeter jets we define a discriminant, the JVF which measures the probability that a jet originated from a particular vertex. The fractional uncertainties of the JVF requirement are obtained by varying the cut from 0.5 up and down 0.1 to 0.4 and 0.6, respectively, and reweighting as a function of the leading jet momentum. The size of the uncertainty is about 1 %.

BCH Cleaning "BCH cleaning" removes the effect from non-functioning modules in the calorimeter when calculating the energy deposits. The systematic uncertainties for BCH cleaning have been derived as recommended in [3]. The BCH inefficiencies are flavor dependent and separate upward/downward fluctuations are calculated. The systematic effect of BCH cleaning is negligible.

Luminosity The uncertainty on the integrated luminosity for the data sample is 2.8 %. It is derived using the same methodology as that detailed in [4].

Cross-Section Uncertainties We assume the following uncertainties of cross-sections for the background samples:

The uncertainty of the $t\bar{t}$ production cross-section is assumed to be 7 %, following [5], which is consistent with the ATLAS measurement of top quark pair production [6]. The single top contribution is very small and for simplicity we use the same uncertainty as for $t\bar{t}$, which is consistent with the theoretical uncertainties [7]. The same procedure is used by the top group [8].

For the W + jets we use 20 % for its cross-section uncertainty from the recent ATLAS measurement of W + jets production with b-jets [9]: this also fully covers the largest difference (estimated to be $\sim$8 %) between prediction and data in the dedicated CR as listed in Table 10.2. Distributions shown in Sect. 10.2.1 show that MC samples describe the efficiencies and shape in W + jets sufficiently well.

For the diboson processes, initially we used the uncertainty figures from recent ATLAS measurements of each of the three processes. For WW production we use the uncertainty of the measured ATLAS cross-section of 9 % [10]. This is conservative compared to the smaller theory cross-section of 4.5 %. Similarly for ZZ production 12 % (theory: 4 %) [11] and for WZ production we assume 9 % (theory: 6 %) [12]. The weighted total is calculated to be 6.4 % for the Diboson processes combined, where the different processes are weighted by their individual contributions after requiring at least two b-jets (of the total diboson yield, 2 % are WW, 58 % are WZ, and 40 % are ZZ), and the systematics are treated as uncorrelated. As the ATLAS measurements are for inclusive processes, while this analysis probes certain slices of phase space with large E_T^{miss} where the cross-section uncertainty on diboson processes can be much larger, we adopt the figures from the most recent ATLAS "mono-jet" analysis [13], which provides a linear extrapolation of cross-section uncertainty based on the comparisons of truth level MC samples with shifted systematics for diboson processes in signal regions with jets and large E_T^{miss}. The uncertainty on diboson cross-section ($WW + WZ + ZZ$) is 20 % at $E_T^{miss} > 150$ GeV, 22 % at $E_T^{miss} > 200$ GeV, 26 % at $E_T^{miss} > 300$ GeV, and 30 % at $E_T^{miss} > 400$ GeV. Due to the fact that diboson process is a relatively minor background at lower E_T^{miss}, and has very limited statistics of around or below 1 event at higher E_T^{miss} where statistical uncertainties dominate, the revision of the diboson cross-section uncertainty brings minimal change to the final limit results at 1 % or less. Tables 11.7, 11.8, 11.9, and 11.10 reflect the revised diboson cross-section uncertainty figures, as is used in the final results in Chap. 12.

PDF and α_s Uncertainties The systematic uncertainty due to the choice of parton distribution functions (PDFs) is determined by using the uncertainty eigenvectors provided for multiple PDF sets per the PDF4LHC prescription [14]. For the simulated background processes, the uncertainty due to variations in PDFs, namely MSTW2008 NNLO [15, 16], CT10 NNLO [17, 18], and NNPDF2.3 [19] PDF sets and parton shower models are approximately 6 % for top, 5 % for V+jets, and 5.9 % for diboson. In particular, the uncertainty on top is calculated by combining in quadrature the effect of different showering models (2.4 %), and with/without $t\bar{t}$ p_T reweighting (5.5 %), yielding 6 % consistent with prior analysis. As the

$Z \to \nu\bar{\nu}$ estimation is data-driven, we assume no additional PDF uncertainties, and the theoretical uncertainties are absorbed in the uncertainty on the transfer function, as detailed in Sect. 11.3. PDF uncertainties for signal samples are discussed in detail in the following section, Sect. 11.2.

11.2 Signal Theoretical Uncertainties

The signal samples from MC simulation are produced at leading order (LO). An estimated value of 10 % is used as the uncertainty on the signal cross-section from NLO corrections [20].

The systematic uncertainty due to the choice of parton distribution functions is determined by using the uncertainty eigenvectors provided for multiple PDF sets per the PDF4LHC prescription [14]. For the signal samples, the PDF uncertainty is estimated as the maximum difference in detector acceptance when using different variations in the MSTW2008 LO [15] and NNPDF2.1 [19] PDF sets, leading to an uncertainty of ∼4–8 % for the Z'-2HDM model. The process of obtaining the PDF uncertainties for the signal samples and the values for each variation is detailed below.

The signal samples are produced with the central value of PDF set MSTW2008 LO: following the PDF4LHC prescription, the PDF uncertainties are calculated using the uncertainty eigenvectors ("error sets") from the MSTW2008LO and NNPDF2.1 PDFs. In addition to the central value, MSTW2008LO has 40 error sets; and NNPDF2.1 has 100 error sets. The events are reweighted with the values from either the nominal value or the one from each error set, and the variation in detector acceptance when requiring at least two b-tagged jets in the final state is calculated for each set.

The PDF uncertainty of an observable can be evaluated in three ways: intra-PDF uncertainty, which is the uncertainty within a given PDF set; inter-PDF uncertainty, which is the variation when switching from one PDF (set) to another PDF; and full-PDF uncertainty as the combination of the inter- and intra-PDF uncertainty.

First, to calculate the intra-PDF uncertainty, the methods are conceptually different for the two families of PDFs. For MSTW, there is one central PDF set and 40 error sets, from which the uncertainties are constructed following the asymmetric Hessian procedure [21]. The asymmetric errors are calculated using the following formulae:

$$\Delta X^+ = \sqrt{\sum_i \max(0, (X_{2i} - X_0), (X_{2i-1} - X_0))^2} \quad , \quad i = 1 \ldots N/2$$

$$\Delta X^- = \sqrt{\sum_i \max(0, (X_0 - X_{2i}), (X_0 - X_{2i-1}))^2} \quad , \quad i = 1 \ldots N/2$$

Here, X_0 is the central value, X_{2i} (X_{2i-1}) corresponds to the $2i$-th ($2i-1$) error set, and N is the total number of error sets, which is 40 here.

The NNPDF sets, on the other hand, are 100 independent sets. The central value is given by the mean value (X_0) of this ensemble and the (symmetric) error by its standard deviation:

$$\Delta X = \sqrt{\frac{1}{100-1} \sum_i (X_i - X_0)^2} \quad , \quad i = 1 \ldots 100$$

The final PDF uncertainty is obtained per PDFLHC4 recommendation by taking the envelope of the variations and uncertainties: the extremum (min and max) of all variations is taken and half of the interval is the absolute uncertainty, i.e.:

$$\Delta X_{\text{env}} = \frac{1}{2} \cdot [\max(X_0^{\text{NNPDF}} + \Delta X, X_0^{\text{MSTW}} + \Delta X^+) - \min(X_0^{\text{NNPDF}} - \Delta X, X_0^{\text{MSTW}} - \Delta X^-)]$$

To obtain the relative uncertainty, ΔX_{env} is divided by the default value of the signal samples, in this case X_0^{MSTW}.

The variation of the acceptance at the two b-tagged jets selection stage due to using a different PDF set is estimated by reweighting the original sample to the alternate PDF and comparing the weighted acceptance to the original. This reweighting is done with the help of the PDFTool [22]. It assigns a weight to each event based on the scale of the event (Q^2), the momentum fractions (x_1, x_2) and types of the two interacting partons, and the original and alternate PDF to be used. The tool uses the LHAPDF library of PDF sets (see also reference [23]). The sets used are 21,000 (MSTW2008lo68cl) and 192,800 (NNPDF21). Hence, the numbers quoted are the uncertainties at 68 % confidence level.

Figure 11.1 shows the variations in acceptance at the two b-tagged jets selection stage for error sets in MSTW2008LO (left) and NNPDF2.1 (right) for $Z' \to A^0 h$ signal sample with $m_{Z'} = 1000\,\text{GeV}$, $m_{A^0} = 300\,\text{GeV}$. The envelope of the variations across the error sets is taken as the PDF uncertainty for each signal sample, as tabulated in Table 11.1 for $Z' \to A^0 h$, and Table 11.2 for $Z' \to Zh$.

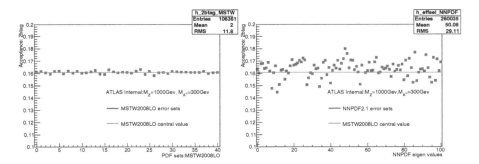

Fig. 11.1 Variations in acceptance at the two b-tagged jets stage for error sets in MSTW2008LO (*left*) and NNPDF2.1 (*right*) for $Z' \to A^0 h$ signal sample with $m_{Z'} = 1000\,\text{GeV}$, $m_{A^0} = 300\,\text{GeV}$

11.2 Signal Theoretical Uncertainties

Table 11.1 PDF uncertainty calculated in terms of acceptance at the two b-tagged jets selection stage for the $Z' \to A^0 h$ signal, varying $M_{Z'}$ and M_{A^0}

$M_{Z'}$ (GeV)	M_{A^0} (GeV)	MSTW2008LO (%)	NNPDF2.1 (%)	Uncertainty (%)
600	300	$9.3^{+0.2}_{-0.2}$	10.2 ± 0.3	7.3
600	400	$4.7^{+0.1}_{-0.1}$	5.1 ± 0.2	7.1
800	300	$15.0^{+0.3}_{-0.4}$	15.8 ± 0.6	5.8
800	400	$14.0^{+0.3}_{-0.3}$	14.8 ± 0.5	5.9
800	500	$11.0^{+0.3}_{-0.3}$	11.6 ± 0.4	5.9
800	600	$7.5^{+0.2}_{-0.2}$	7.8 ± 0.3	5.6
1000	300	$16.1^{+0.4}_{-0.5}$	16.3 ± 0.7	4.4
1000	400	$16.0^{+0.4}_{-0.5}$	16.4 ± 0.7	4.7
1000	500	$15.6^{+0.4}_{-0.4}$	15.9 ± 0.7	4.6
1000	600	$14.7^{+0.4}_{-0.4}$	14.9 ± 0.6	4.4
1000	700	$11.8^{+0.3}_{-0.3}$	12.0 ± 0.5	4.5
1000	800	$10.1^{+0.3}_{-0.3}$	10.2 ± 0.4	4.4
1200	300	$14.8^{+0.5}_{-0.5}$	14.6 ± 0.7	4.8
1200	400	$15.4^{+0.5}_{-0.5}$	15.2 ± 0.7	4.8
1200	500	$15.7^{+0.5}_{-0.5}$	15.3 ± 0.8	5.1
1200	600	$16.1^{+0.5}_{-0.5}$	15.8 ± 0.8	4.8
1200	700	$15.6^{+0.5}_{-0.5}$	15.3 ± 0.8	4.9
1200	800	$14.5^{+0.5}_{-0.5}$	14.1 ± 0.7	5.1
1400	300	$10.1^{+0.4}_{-0.4}$	9.6 ± 0.5	6.6
1400	400	$11.2^{+0.4}_{-0.4}$	10.7 ± 0.6	6.8
1400	500	$13.0^{+0.5}_{-0.5}$	12.4 ± 0.7	6.7
1400	600	$14.5^{+0.5}_{-0.5}$	13.7 ± 0.8	7.0
1400	700	$16.1^{+0.6}_{-0.6}$	15.3 ± 0.8	6.9
1400	800	$15.8^{+0.6}_{-0.6}$	15.0 ± 0.8	7.0

The columns from left to right describe $M_{Z'}$, M_{A^0}, $X_0^{\text{MSTW}} + \Delta X^+ - \Delta X^-$, $X_0^{\text{NNPDF}} \pm \Delta X$, and the final PDF uncertainty calculated with the envelope method. The acceptance and uncertainty numbers are in percentage

Table 11.2 PDF uncertainty calculated in terms of acceptance at two b-tag stage for the $Z' \to Zh$ process, varying $M_{Z'}$

$M_{Z'}$ (GeV)	MSTW2008LO (%)	NNPDF2.1 (%)	Uncertainty (%)
600	$12.9^{+0.2}_{-0.3}$	14.1 ± 0.4	7.3
800	$15.0^{+0.3}_{-0.4}$	15.8 ± 0.6	5.9
1000	$15.0^{+0.4}_{-0.4}$	15.4 ± 0.6	4.7
1200	$13.5^{+0.4}_{-0.4}$	13.3 ± 0.6	4.7
1400	$8.2^{+0.3}_{-0.3}$	7.9 ± 0.4	6.2

The columns from left to right describe $M_{Z'}$, $X_0^{\text{MSTW}} + \Delta X^+ - \Delta X^-$, $X_0^{\text{NNPDF}} \pm \Delta X$, and the final PDF uncertainty calculated with the envelope method. The acceptance and uncertainty numbers are in percentage

11.3 $Z(\to \nu\bar{\nu})$ + Jets Background Systematic Uncertainties

Due to the data-driven methods to estimate $Z(\to \nu\bar{\nu})$+jets background as described in Sect. 10.3.2, the systematic uncertainties associated with it come from two sources, the uncertainties from the fitting of the transfer functions, and the experimental uncertainties as described in the previous section that propagate through the fitting and reweighting process due to the use of MC samples in the derivation of the transfer functions.

The final $Z(\to \nu\bar{\nu})$ + jets estimation comes from two parts: when E_T^{miss} is below 200 GeV, the estimation is reweighted from $Z \to \mu\mu$, where other background sources (top, W+jets, diboson) that pass the two muon selection are subtracted from 2μ data; when E_T^{miss} is above 200 GeV, the estimation is reweighted from γ + jets data, where the purity is very high (around 99% after b-tagging). The two methods yield two distinct sets of systematics. In our final signal selection, we adopt a sliding E_T^{miss} cut going from 150 to 400 GeV for signal samples with different Z' and A^0 masses: the number of $Z \to \nu\bar{\nu}$ events coming from the two different methods is listed in Table 11.3.

For signal samples using the E_T^{miss} above 150 GeV cut, the $Z(\to \nu\bar{\nu})$ + jets estimation comes about half from reweighted $Z(\to \mu\mu)$ + jets events and half from γ + jets events, and the systematics associated with both methods are used in the final combined total. For signal samples using the E_T^{miss} above 200 GeV cut or above, the $Z(\to \nu\bar{\nu})$ +jets estimation comes only from γ +jets events, and the systematics associated with the γ + jets method is used.

The sources of systematic errors from the fit of the transfer function include the function used for the fit, the shape of the fit, selection stage used for the fit, the statistical error of the fit, and variations on the E_T^{miss} range of the fit.

In the E_T^{miss} below 200 GeV region where $Z \to \nu\bar{\nu}$ is estimated from $Z \to \mu\mu$, the transfer function is fit with a four-variable polynomial "pol4". To estimate the systematic effects, alternative polynomials "pol3" and "pol5" are used to fit the transfer function. The different polynomials yield a similar χ^2/dof, as tabulated in Table 11.4, so "pol4" is used for the final results, and the variations from the

Table 11.3 $Z(\to \nu\bar{\nu})$ + jets yield in the signal region (SR)

	$Z(\to \nu\nu)$ total	From $Z \to \mu\mu$	From γ + jets
At least two b-btag 60% eff.	520 ± 112	453 ± 117	67 ± 4
90 GeV $\leq m_{bb} \leq$ 150 GeV	199 ± 79	176 ± 79	23 ± 3
$\Delta R(j_0, j_1) < 1.5$	94 ± 44	73 ± 44	21 ± 2
$E_T^{miss} \geq 150$ GeV	47.74 ± 25.85	26.56 ± 25.73	21.18 ± 2.49
$E_T^{miss} \geq 200$ GeV	21.18 ± 2.49	–	21.18 ± 2.49
$E_T^{miss} \geq 300$ GeV	2.87 ± 0.96	–	2.87 ± 0.96
$E_T^{miss} \geq 400$ GeV	0.33 ± 0.33	–	0.33 ± 0.33

Final four rows show results in SR with sliding E_T^{miss} cut. The uncertainties show statistical errors only

11.3 $Z(\to \nu\bar{\nu})$ + Jets Background Systematic Uncertainties

Table 11.4 χ^2 variations from different transfer functions used to fit $Z \to \nu\bar{\nu}$ from $Z \to \mu\mu$

Fit function	χ^2/dof	dof	χ^2	$\delta\chi^2$
pol3	1.52	173	263.7	0.4
pol4	1.51	174	263.3	–
pol5	1.49	175	261.6	-1.7

"pol4" is used for the final results

alternative polynomials are taken into account as systematics. In the E_T^{miss} above 200 GeV region where γ + jets is used to estimate $Z \to \nu\bar{\nu}$, a physics-motivated function, Eq. (10.4), is used: as the fit parameters are very close to theoretical calculations, we assume no additional systematic errors from this fit function itself.

To estimate the systematic effect of different selection stages, transfer functions are derived at such stages before or after cuts on jet p_T and multiplicity are applied (Fig. 10.15 for estimation from $Z \to \mu\mu$, Fig. 10.23 for estimation from γ + jets): the transfer function from the stage after the cut on leading jet p_T is used for final results, and variations from functions derived from other selection stages are taken into account as systematics. For the fit range, the minimum and maximum E_T^{miss} of the fit are varied independently of each other: no change in fit parameter was observed by this variation, so we deduce that there is no additional systematics associated with fit range. Finally, the statistical error from the fit is accounted for by varying each of the fit parameters by one standard deviation in either direction, giving ten new curves in the $Z \to \mu\mu$ method part, and four new curves in the γ + jets method part. The error matrix shows that the parameters are almost fully correlated, so the envelope of the variation from each of the new curves is taken into account as systematics.

For each of the aforementioned variations, the resulting transfer function is applied in the reweighting process instead of the original function, and the difference in $Z \to \nu\bar{\nu}$ estimation after the $2 - b$-tagged jets requirement is taken as systematic uncertainty.

While the fit functions describe the shape ratios of $Z \to \nu\bar{\nu}$ E_T^{miss} distributions and that of $Z \to \mu\mu$ or γ + jets sufficiently well, we take the small fluctuations in the shape of the ratio and that of the function as an additional systematic called "fit shape." As the transfer functions are calculated from MC samples alone, for E_T^{miss} between 150 and 200 GeV where $Z \to \nu\bar{\nu}$ is estimated from $Z \to \mu\mu$, the integrated event count of $Z \to \nu\bar{\nu}$ MC in said E_T^{miss} range is compared with that of $Z \to \mu\mu$ MC reweighted by the transfer function, and the deviation (1 %) is taken into account as the systematic "fit shape"; similarly for E_T^{miss} above 200 GeV where $Z \to \nu\bar{\nu}$ is estimated from γ + jets, we compare the integrated event count of $Z \to \nu\bar{\nu}$ MC, and that of γ + jets MC reweighted by the transfer function, in three regions, E_T^{miss} above 200, 300, and 400 GeV, respectively, and take the envelope of the deviation (2.5 %) as the additional systematic.

As γ + jets MC is used for the derivation of the transfer function in E_T^{miss} above 200 GeV region, there is an additional source of systematics uncertainty in the transfer function from photon ID/reconstruction efficiency, isolation, energy scale,

Table 11.5 Summary of systematic uncertainties in percent from fitting of transfer function in $Z \to \nu\bar{\nu}$ data-driven estimation

$Z \to \nu\bar{\nu}$ estimation	Fit function	Fit error	Fit range in E_T^{miss}	Fit stage	Fit shape	Photon-sys	Total
$Z \to \mu\mu$ method	0.5	8	0	0.2	1	–	8.1
γ + jets method	0	0.2	0	4.6	2.5	4.0	6.9

and resolution. The uncertainty on photon ID/reconstruction efficiency is calculated by varying the "fudge factors" (correction factors computed by comparing all shower shapes observed in data and MC samples) up and down, recalculating the transfer function and comparing the difference in final $Z \to \nu\bar{\nu}$ yield after reweighting, which gives 0.5%. Similarly, the photon energy scale and resolution uncertainties are calculated to be around 0.1% or below. For the photon isolation uncertainty, we take an overall conservative estimate of 4%, following the studies of the 8 TeV "Mono-photon" search [24] which uses the same tight photon isolation requirements as this analysis. The overall photon-related systematic uncertainty is hence 4.0% by combining the aforementioned sources in quadrature.

The uncertainties associated with each of the sources are tabulated in Table 11.5. Uncertainties from the six sources are added in quadrature for the total uncertainty in transfer function alone, listed in the final column.

In addition to the fitting of the transfer function itself, as the transfer function is derived from $Z \to \nu\bar{\nu}$ MC sample divided by either $Z \to \mu\mu$ MC sample or γ + jets MC sample, detector systematics in Sect. 11.1 are propagated to each of the MC samples to calculate the respective transfer functions associated with each variation. As the effects of the systematics tend to cancel out in the dividing process, the change in the transfer function itself due to detector systematics is small. In E_T^{miss} above 200 GeV region, reweighting is performed on γ + jets data alone, so no additional systematic uncertainties are introduced in the reweighting process. In E_T^{miss} below 200 GeV region, reweighting is performed on 2μ data, with other backgrounds (top, W+jets, diboson) subtracted from MC. In this case, the detector systematic variations are applied to the MC samples used in subtraction as well, which means for each of the detector systematics, the uncertainty in $Z \to \nu\bar{\nu}$ is anti-correlated from the rest of the background processes (Table 11.6).

The final variation in $Z \to \nu\bar{\nu}$ estimation associated with each detector systematics is considered experimental uncertainty and tabulated for the individual variations in the next section in Tables 11.7, 11.8, 11.9, and 11.10. The uncertainty from the transfer function fitting itself as tabulated in Table 11.5 is considered theoretical uncertainty in Tables 11.7, 11.8, 11.9, and 11.10. Finally, we take the effect of the 0.9 scale factor derived from the $Z \to \nu\bar{\nu}$ control regions (Table 10.10), yielding a 10% cross-section uncertainty for the $Z \to \nu\bar{\nu}$ estimation.

11.4 Systematic Uncertainties in Signal Region

Table 11.6 Summary of systematic uncertainty in percent for all backgrounds combined and signal samples in the resolved and boosted channels

	Z'-2HDM	Total background
b-tagging	14	6–10
JES	2.4	1.8–2.8
JER	0.6	3.5–5.4
JVF	0.7	0.5–0.9
E_T^{miss} resolution/scale	0.0	< 0.2
Pile-up	0.3	0.1
Cross-section	10	6.0–11
PDF and α_s	4.4–7.3	2.9
$Z(\nu\bar{\nu})$ transfer function	–	1.4–2.7
Total syst.	18–19	10–16

The first column lists the main sources of systematic uncertainty, where the acronym JES refers to the jet energy scale, JER the jet energy resolution, and JVF the JVF. The uncertainty figures listed for "b-tagging" combine the uncertainty from both b-tagging efficiency and mistag rates. The uncertainty ranges in "Total Background" reflect the shift in value with increasing E_T^{miss} threshold in the final signal region. The uncertainties for "$Z(\nu\bar{\nu})$ transfer function" take into account the fractional weight of the $Z(\nu\bar{\nu})$ process in total background, which differs per analysis channel and E_T^{miss} threshold. Most of the systematic uncertainties on the signal models vary little across the parameter space in this analysis, with the exception of signal PDF and α_s, and pile-up uncertainty; hence the ranges of values are shown

11.4 Systematic Uncertainties in Signal Region

All systematic uncertainties per process and their source are listed in Table 11.7 for E_T^{miss} above 150 GeV, Table 11.8 for E_T^{miss} above 200 GeV, Table 11.9 for E_T^{miss} above 300 GeV, and Table 11.10 for E_T^{miss} above 400 GeV. In Table 11.7, the systematics for $Z \to \nu\bar{\nu}$ process are split into two columns for estimation from $Z \to \mu\mu$ and γ + jets, respectively, and in Tables 11.8, 11.9, 11.10, and $Z \to \nu\bar{\nu}$ systematics consist of only one column when it is estimated from γ + jets only, as explained in the previous section. The systematics for the rest of the background processes should not be affected by the E_T^{miss} requirement, hence for higher statistics and a more accurate calculation, they are obtained at E_T^{miss} above 100 GeV, and the same numbers are applied in all four tables. The final column in each table shows the systematics for all background processes combined, with each background process' fractional contribution applied as weight in the calculation.

We can see here that the main detector systematics are b-tagging (up/down), JES (up/down), and JER. For b-tagging systematics, the diboson, V + jets, and top quark processes are treated as fully correlated, and the $Z \to \nu\bar{\nu}$ process is treated as anti-correlated. This leads to a significant reduction in b-tagging systematics, especially in Table 11.7, which is reasonable and expected, as in E_T^{miss} above 100 GeV region,

Table 11.7 Summary of systematic uncertainties in percent for all backgrounds and signal sample in the signal region with $E_T^{miss} > 150\,\text{GeV}$

$E_T^{miss} >$ 150 GeV	DM+H($b\bar{b}$)	Diboson	$V+$jets	Top	$Z \to \nu\bar{\nu}\,(\mu\mu)$	$Z \to \nu\bar{\nu}\,(\gamma)$	Tot.Bkgd.
BCH down	−0.5	−0.2	0.0	−0.5	0.5	0	−0.2
BCH up	0.4	0.3	0.1	0.5	0.5	0	0.2
b-tagging down	−13.4	−11.6	−11.7	−13.1	12.8	−0.6	−5.3
b-tagging up	14.3	12.3	12.6	14	−12.5	0.4	5.9
c-tagging down	0.0	−1.1	−2.7	−3.4	3.4	0.0	−1.3
c-tagging up	0.0	1.2	2.6	3.6	3.6	0.0	1.4
Light flavor tag. down	0.0	0.0	−1.6	−0.3	1.6	0.0	−0.4
Light flavor tag. up	0.0	0.0	1.3	0.6	1.3	0.0	0.35
JES down	2.2	−1.8	−3.4	7.7	5.1	1	2.7
JES up	−2.5	−4.7	−5.8	−7.7	−5.1	−0.6	−2.9
JER	−0.6	−6.8	−9.4	−15.4	6.6	−0.2	−5.4
JVF down	−0.5	−0.9	−0.8	−2.6	2.6	0.0	−1.0
JVF up	0.8	0.2	−0.2	2.1	2.1	0.0	0.8
E_T^{miss} energy res	0.0	0.0	−0.4	−0.4	0.4	0.0	−0.16
E_T^{miss} energy scale down	0.0	0.0	0.0	0.0	0.0	0.0	0
E_T^{miss} energy scale up	0.0	0.0	0.1	0.1	0.1	0.0	0.04
Pile-Up-Rw down	0.2	−0.3	0.4	0.0	0.4	0.0	−0.1
Pile-Up-Rw up	−0.4	0.5	−0.2	0.0	0.5	0.0	0.14
xsec	10.0	20.0	20.0	7.0	10.0	10.0	6.0
PDF & α_s	Table 11.1	5.9	5.0	6.0	0.0	0.0	2.9
TF ($Z \to \nu\bar{\nu}$)					8.1	6.9	
Exp. $\sqrt{\sum \sigma_i^2}$ up	14.5	13.2	14.2	16.5	16	0.7	6.7
Exp. $\sqrt{\sum \sigma_i^2}$ down	−13.6	−13.6	−15.7	−22.1	−14.2	−1.2	−8.3
Theo. uncert.		20.9	20.6	9.2	12.9	12.2	6.7

As $ZH(bb)$ events with E_T^{miss} below 200 GeV are included in the $Z \to \nu\bar{\nu}$ estimation reweighted from $Z \to \mu\mu$ events, and the rest of the $VH(bb)$ events contribute about 3 % of the total background here, its effect on total combined background is negligible and hence omitted from this table

11.4 Systematic Uncertainties in Signal Region

Table 11.8 Summary of systematic uncertainties in percent for all backgrounds and signal sample in the signal region with $E_T^{miss} > 200\,\text{GeV}$

$E_T^{miss} > 200\,\text{GeV}$	DM+H($b\bar{b}$)	Diboson	V + jets	Top	$Z \to \nu\nu$	$VH(bb)$	Total Bkgd.
BCH down	−0.5	−0.2	0.0	−0.5	0	0	−0.15
BCH up	0.4	0.3	0.1	0.5	0	−0.5	0.16
b-tagging down	−13.4	−11.6	−11.7	−13.1	−0.6	−12.3	−8.0
b-tagging up	14.3	12.3	12.6	14	0.4	12.5	8.6
c-tagging down	0.0	−1.1	−2.7	−3.4	0.0	0.1	−1.1
c-tagging up	0.0	1.2	2.6	3.6	0.0	−0.3	1.1
Light flavor tag. down	0.0	0.0	−1.6	−0.3	0.0	−0.1	−0.18
Light flavor tag. up	0.0	0.0	1.3	0.6	0.0	−0.3	0.21
JES down	2.2	−1.8	−3.4	7.7	1	−2.0	2.3
JES up	−2.5	−4.7	−5.8	−7.7	−0.6	2.0	−2.5
JER	−0.6	−6.8	−9.4	−15.4	−0.2	−2.7	−4.7
JVF down	−0.5	−0.9	−0.8	−2.6	0.0	−0.5	−0.7
JVF up	0.8	0.2	−0.2	2.1	0.0	0.4	0.6
E_T^{miss} energy res	0.0	0.0	−0.4	−0.4	0.0	−0.2	−0.1
E_T^{miss} energy scale down	0.0	0.0	0.0	0.0	0.0	0.2	−0.02
E_T^{miss} energy scale up	0.0	0.0	0.1	0.1	0.0	−0.2	0.04
Pile-Up-Rw down	0.2	−0.3	0.4	0.0	0.0	0.2	−0.07
Pile-Up-Rw up	−0.4	0.5	−0.2	0.0	0.0	−0.6	0.1
xsec	10.0	22.0	20.0	7.0	10.0	3.1	6.2
PDF & α_s	Table 11.1	5.9	5.0	6.0	0.0	2.5	2.9
TF ($Z \to \nu\bar{\nu}$)					6.9		
Exp. $\sqrt{\sum \sigma_i^2}$ up	14.5	13.2	14.2	16.5	0.7	12.7	9.0
Exp. $\sqrt{\sum \sigma_i^2}$ down	−13.6	−13.6	−15.7	−22.1	−1.2	−12.8	−9.7
Theo. uncert.		22.8	20.6	9.2	12.2	4.0	6.8

Table 11.9 Summary of systematic uncertainties in percent for all backgrounds and signal sample in the signal region with $E_T^{miss} > 300$ GeV

$E_T^{miss} > 300$ GeV	DM+H($b\bar{b}$)	Diboson	V + jets	Top	$Z \to \nu\nu$	$VH(bb)$	Total Bkgd.
BCH down	−0.5	−0.2	0.0	−0.5	0	0	−0.1
BCH up	0.4	0.3	0.1	0.5	0	−0.5	0.15
b-tagging down	−13.4	−11.6	−11.7	−13.1	−0.6	−12.3	−8.4
b-tagging up	14.3	12.3	12.6	14	0.4	12.5	9
c-tagging down	0.0	−1.1	−2.7	−3.4	0.0	0.1	−0.76
c-tagging up	0.0	1.2	2.6	3.6	0.0	−0.3	0.8
Light flavor tag. down	0.0	0.0	−1.6	−0.3	0.0	−0.1	−0.2
Light flavor tag. up	0.0	0.0	1.3	0.6	0.0	−0.3	0.2
JES down	2.2	−1.8	−3.4	7.7	1	−2.0	1.5
JES up	−2.5	−4.7	−5.8	−7.7	−0.6	2.0	−2.1
JER	−0.6	−6.8	−9.4	−15.4	−0.2	−2.7	−3.5
JVF down	−0.5	−0.9	−0.8	−2.6	0.0	−0.5	−0.5
JVF up	0.8	0.2	−0.2	2.1	0.0	0.4	0.4
E_T^{miss} energy res	0.0	0.0	−0.4	−0.4	0.0	−0.2	−0.7
E_T^{miss} energy scale down	0.0	0.0	0.0	0.0	0.0	0.2	−0.02
E_T^{miss} energy scale up	0.0	0.0	0.1	0.1	0.0	−0.2	0.03
Pile-Up-Rw down	0.2	−0.3	0.4	0.0	0.0	0.2	−0.11
Pile-Up-Rw up	−0.4	0.5	−0.2	0.0	0.0	−0.6	0.17
xsec	10.0	26.0	20.0	7.0	10.0	3.1	8.8
PDF & α_s	Table 11.1	5.9	5.0	6.0	0.0	2.5	2.7
TF ($Z \to \nu\bar{\nu}$)					6.9		
Exp. $\sqrt{\sum \sigma_i^2}$ up	14.5	13.2	14.2	16.5	0.7	12.7	9.2
Exp. $\sqrt{\sum \sigma_i^2}$ down	−13.6	−13.6	−15.7	−22.1	−1.2	−12.8	−9.4
Theo. uncert.		26.7	20.6	9.2	12.2	4.0	9.2

$Z \to \nu\bar{\nu}$ contributes about half of the background events: its detector systematics, as explained from the previous section, mainly comes when $Z \to \nu\bar{\nu}$ is estimated from $Z \to \mu\mu$ and other MC backgrounds are subtracted from 2μ data in the reweighting process, leading to the anti-correlation and canceling out of the systematic variations. The detector systematics associated with $Z \to \nu\bar{\nu}$ estimated from γ + jets is minimal as the reweighting is on γ + jets data only. For JES (up/down) and JER, considering that the numerous variables used in the calculation would differ significantly for each background process, these uncertainties are treated as uncorrelated in calculating the combined total. The rest of the detector systematics yield minimal fluctuations, and are treated as uncorrelated among the different background processes.

The cross-section uncertainties are treated as uncorrelated among the background processes. PDF & α_s uncertainties, as well as the uncertainty from the fit of the transfer function alone in $Z \to \nu\bar{\nu}$ estimation (Table 11.5), are considered as theoretical uncertainties and treated as uncorrelated. The "theoretical uncertainties"

11.4 Systematic Uncertainties in Signal Region

Table 11.10 Summary of systematic uncertainties in percent for all backgrounds and signal sample in the signal region with $E_T^{miss} > 400\,\text{GeV}$

$E_T^{miss} > 400\,\text{GeV}$	DM+H($b\bar{b}$)	Diboson	V + jets	Top	$Z \to \nu\nu$	VH(bb)	Total Bkgd.
BCH down	−0.5	−0.2	0.0	−0.5	0	0	−0.11
BCH up	0.4	0.3	0.1	0.5	0	−0.5	0.13
b-tagging down	−13.4	−11.6	−11.7	−13.1	−0.6	−12.3	−9.8
b-tagging up	14.3	12.3	12.6	14	0.4	12.5	10.2
c-tagging down	0.0	−1.1	−2.7	−3.4	0.0	0.1	−0.8
c-tagging up	0.0	1.2	2.6	3.6	0.0	−0.3	0.8
Light flavor tag. down	0.0	0.0	−1.6	−0.3	0.0	−0.1	−0.25
Light flavor tag. up	0.0	0.0	1.3	0.6	0.0	−0.3	0.23
JES down	2.2	−1.8	−3.4	7.7	1	−2.0	1.4
JES up	−2.5	−4.7	−5.8	−7.7	−0.6	2.0	−2.2
JER	−0.6	−6.8	−9.4	−15.4	−0.2	−2.7	−3.7
JVF down	−0.5	−0.9	−0.8	−2.6	0.0	−0.5	−0.5
JVF up	0.8	0.2	−0.2	2.1	0.0	0.4	0.35
E_T^{miss} energy res	0.0	0.0	−0.4	−0.4	0.0	−0.2	−0.09
E_T^{miss} energy scale down	0.0	0.0	0.0	0.0	0.0	0.2	−0.03
E_T^{miss} energy scale up	0.0	0.0	0.1	0.1	0.0	−0.2	0.04
Pile-Up-Rw down	0.2	−0.3	0.4	0.0	0.0	0.2	−0.12
Pile-Up-Rw up	−0.4	0.5	−0.2	0.0	0.0	−0.6	0.17
xsec	10.0	30.0	20.0	7.0	10.0	3.1	11.3
PDF & α_s	Table 11.1	5.9	5.0	6.0	0.0	2.5	2.6
TF ($Z \to \nu\bar{\nu}$)					6.9		
Exp. $\sqrt{\sum \sigma_i^2}$ up	14.5	13.2	14.2	16.5	0.7	12.7	10.3
Exp. $\sqrt{\sum \sigma_i^2}$ down	−13.6	−13.6	−15.7	−22.1	−1.2	−12.8	−10.7
Theo. uncert.		30.6	20.6	9.2	12.2	4.0	11.6

are listed in separate (bottom two) rows in Tables 11.7, 11.8, 11.9, and 11.10: as there is no additional PDF & α_s uncertainty from $Z \to \nu\bar{\nu}$ data-driven estimation, the uncertainty on transfer function itself is added in quadrature with the rest of the PDF & α_s uncertainties for the final number as the theoretical uncertainty for the combined background, listed in the rightmost column of the tables on the bottom above the two rows for $\sqrt{\sum \sigma_i^2}$.

For the final combined background uncertainties from each systematics source, uncorrelated uncertainties from each background process are added in quadrature with the relative weight of each process, and (anti-)correlated uncertainties are added in sum with the relative weight of each process applied. Overall, the systematic uncertainty on the estimated background is calculated to be between 10 and 16 %, depending on the final E_T^{miss} requirement in the signal region.

The uncertainty for the DM+H($b\bar{b}$) signal listed here is derived from the $M_{Z'} = 1000\,\text{GeV}$, $M_{A^0} = 300\,\text{GeV}$, and $M_{\text{DM}} = 100\,\text{GeV}$ sample. Variations on each systematic uncertainty across different signal samples are found to be very small and generally below 1 %.

In the limit setting, the detector systematics are fully correlated between signal and background; the theoretical uncertainties (PDF & α_s, fit of the transfer function), as well as cross-section uncertainties, are uncorrelated between signal and background. Table 11.6 lists the main sources of systematic uncertainty, and their values for both signals and backgrounds. The values given for the backgrounds are the uncertainties on the total background with the relative weights and correlations of individual background processes taken into account.

References

1. ATLAS Collaboration, Calibration of b-tagging using dileptonic top pair events in a combinatorial likelihood approach with the ATLAS experiment. (ATLAS-CONF-2014-004), Feb (2014)
2. ATLAS Collaboration, Online https://twiki.cern.ch/twiki/bin/view/AtlasProtected/MultijetJESUncertaintyProvider
3. C. Doglioni, Bchcleaningtool (2013). Online https://twiki.cern.ch/twiki/bin/viewauth/AtlasProtected/BCHCleaningTool
4. ATLAS Collaboration, Improved luminosity determination in pp collisions at $\sqrt{s} = 7\,\text{TeV}$ using the ATLAS detector at the LHC. Eur. Phys. J. **C73**, 2518 (2013)
5. S. Moch, P. Uwer, Heavy-quark pair production at two loops in QCD. Nucl. Phys. Proc. Suppl. **183**, 75–80 (2008)
6. ATLAS Collaboration, Measurement of the top pair production cross-section in 8 TeV proton-proton collisions using kinematic information in the lepton+jets final state with ATLAS. Phys. Rev. **D91**, 112013 (2015)
7. N. Kidonakis, Two-loop soft anomalous dimensions for single top quark associated production with a W^- or H^-. Phys. Rev. **D82**, 054018 (2010)
8. Measurement of the inclusive and fiducial cross-section of single top-quark t-channel events in pp collisions at $\sqrt{s} = 8\,\text{TeV}$. Technical Report ATLAS-CONF-2014-007, CERN, Geneva, Mar (2014)
9. G. Aad et al., Measurement of the cross-section for W boson production in association with b-jets in pp collisions at $\sqrt{s} = 7\,\text{TeV}$ with the ATLAS detector. J. High Energy Phys. **1306**, 084 (2013)
10. ATLAS Collaboration, Measurement of W^+W^- production in pp collisions at $\sqrt{s} = 7\,\text{TeV}$ with the ATLAS detector and limits on anomalous WWZ and WWγ couplings. Phys. Rev. **D87**(11), 112001 (2013). [Erratum: Phys. Rev. **D88**(7), 079906 (2013)]
11. ATLAS Collaboration, Measurement of ZZ production in pp collisions at $\sqrt{s} = 7\,\text{TeV}$ and limits on anomalous ZZZ and ZZγ couplings with the ATLAS detector. J. High Energy Phys. **1303**, 128 (2013)
12. ATLAS Collaboration, Measurement of WZ production in proton-proton collisions at $\sqrt{s} = 7\,\text{TeV}$ with the ATLAS detector. Eur. Phys. J. **C72**, 2173 (2012)

13. J. Abdallah, K.A. Assamagan, D. Berge, P. Calfayan, G. Carrillo-Montoya, C. Clement, A.C. Gonzalez, Y. Cheng, J. Gramling, O. Lundberg, P. Mal, M. Martinez, C. Mwewa, J. Pearce, B. Penning, R. Poettgen, R. Rezvani, V. Rossetti, D. Salek, S. Schramm, L. Tompkins, T. Vickey, J. Webster, X. Wu, Search for new phenomena with mono-jet plus missing transverse energy signature in proton-proton collisions at $\sqrt{s}$ = 8 TeV with the ATLAS detector: inclusive analysis. Technical Report ATL-COM-PHYS-2014-1052, CERN, Geneva, Aug (2014)
14. M. Botje et al., The PDF4LHC working group interim recommendations (2011)
15. A.D. Martin, W.J. Stirling, R.S. Thorne, G. Watt, Parton distributions for the LHC. Eur. Phys. J. **C63**, 189–285 (2009)
16. A.D. Martin, W.J. Stirling, R.S. Thorne, G. Watt, Uncertainties on alpha(S) in global PDF analyses and implications for predicted hadronic cross sections. Eur. Phys. J. **C64**, 653–680 (2009)
17. H.-L. Lai, M. Guzzi, J. Huston, Z. Li, P.M. Nadolsky, J. Pumplin, C.P. Yuan, New parton distributions for collider physics. Phys. Rev. **D82**, 074024 (2010)
18. J. Gao, M. Guzzi, J. Huston, H.-L. Lai, Z. Li, P. Nadolsky, J. Pumplin, D. Stump, C.P. Yuan, CT10 next-to-next-to-leading order global analysis of QCD. Phys. Rev. **D89**, 033009 (2014)
19. R.D. Ball et al., Parton distributions with LHC data. Nucl. Phys. **B867**, 244–289 (2013)
20. U. Haisch, F. Kahlhoefer, E. Re, QCD effects in mono-jet searches for dark matter. J. High Energy Phys. **12**, 007 (2013)
21. J.M. Campbell, J.W. Huston, W.J. Stirling, Hard interactions of quarks and gluons: a primer for LHC physics. Rep. Prog. Phys. **70**, 89 (2007)
22. Y. Kataoka, Pdfreweight (2013). Online https://twiki.cern.ch/twiki/bin/viewauth/AtlasProtected/PDFReweight,
23. M.R. Whalley, D. Bourilkov, R.C. Group, The Les Houches accord PDFs (LHAPDF) and LHAGLUE (2005)
24. L. Carminati, D. Cavalli, M.-H. Genest, V. Ippolito, A. Nelson, L. Kashif, M. Perego, C. Pizio, M.G. Ratti, S. Resconi, C. Shimmin, F. Wang, D. Whiteson, M. Wu, S.L. Wu, N Zhou, Search for new phenomena with the ATLAS detector in monophoton events from proton-proton collisions at sqrt(s)=8TeV. Technical Report ATL-COM-PHYS-2014-348, CERN, Geneva, Apr (2014)

Chapter 12
Dark Matter + Higgs($\to b\bar{b}$): Results

This chapter presents the results of this search and the statistical interpretations. For a description of the Z'-2HDM signal model that is used to benchmark this search, see Chap. 7. The signal region selections are defined in Chap. 9. The various background processes are described in Chap. 10.

This chapter is organized as follows: Sect. 12.1.1 presents the numbers of observed data and background in the signal region, and the statistical and systematic uncertainties associated with them; Sect. 12.1.2 lists the final E_T^{miss} requirement optimized for each signal sample in the parameter space probed and the sensitivity estimation, and shows the kinematic distributions in the signal region; the statistical method used to interpret the results is given in Sect. 12.2; the constraints set for the Z'-2HDM model is given in Sect. 12.3; and the model-independent upper limit on the visible cross-section for the DM + $h(\to b\bar{b})$ final state is shown in Sect. 12.4.

12.1 Dark Matter + Higgs($\to b\bar{b}$) Signal Region

As good agreement between data and estimation is reached in the different control region and validation regions described in Chap. 10, the background processes are applied in the signal region as defined in Chap. 9. The systematic uncertainty obtained in Chap. 11 is applied as well. The total expected background and signal events, and their respective uncertainties, are used for final optimization of the sliding E_T^{miss} requirement for individual signals, and compared with data for statistical interpretation of the results.

12.1.1 Event Yield

The observed data, total expected background, and individual background for all selection stages in the signal region are given in Table 12.1.

In addition to the background processes discussed in the previous sections, the standard model Higgs production in association with a vector boson is also included in the signal region as part of the background estimation. As seen in the cutflow table (Table 12.1), $Vh(\to b\bar{b})$ production is a minor background in general, but its contribution becomes non-negligible in the final signal region, especially for E_T^{miss} greater than 200 GeV. Considering our signal region selections, this comes essentially from $Z(\to \nu\nu)+h(\to b\bar{b})$ and $W(\to \ell\nu)+h(\to b\bar{b})$ processes. Since for E_T^{miss} below 200 GeV, the $Z \to \nu\bar{\nu}$ background is estimated from 2μ data (with top, W+jets, and diboson contaminations subtracted from MC), it would have included the $Z(\to \nu\nu)+h(\to b\bar{b})$ events (reweighted from $Z(\to \mu\mu)+h(\to b\bar{b})$ events in the 2μ datastream). Hence, only $Z(\to \nu\nu)+h(\to b\bar{b})$ events with E_T^{miss} above 200 GeV, and $W(\to \ell\nu)+h(\to b\bar{b})$ in the full E_T^{miss} range, are added to the background.

Table 12.2 shows the exclusive (current yield divided by previous yield) and inclusive (current yield divided by initial yield) selection efficiencies for the signal selection.

The predicted number of background events in the signal region for each value of the ascending E_T^{miss} thresholds, along with the number of events observed in the data, including systematic and statistical error on individual and combined background processes, are listed in Table 12.3. For completeness we include in Table 12.3 both the results from the resolved channel, where the Higgs boson is reconstructed as two separate b-jets and is the focus of this thesis, as well as the boosted channel, where the Higgs boson is reconstructed as one large-radius jet using jet substructure techniques, as described in Chap. 6. The overlap between the resolved and boosted channels is found to be small at approximately 15% in signal regions with the same E_T^{miss} requirement. The results from Table 12.3 provide a comprehensive view of the data and background yields in the $E_T^{miss} + h(\to b\bar{b})$ final state, and is used to obtain model-independent upper limits on beyond-the-SM events in Sect. 12.4. The numbers of predicted background events and observed events are consistent within 1σ in five out of the six signal regions. A small excess of 2.2σ is observed in the $E_T^{miss} > 300$ GeV region in the boosted channel. A detailed comparison between the shape of the aforementioned excess and that of the signals in various kinematic distributions leads to the conclusion that the excess is not signal-like and likely due to background fluctuations. This is supported by the calculation of the look-elsewhere effect described in Sect. 12.4.

12.1 Dark Matter + Higgs($\to b\bar{b}$) Signal Region

Table 12.1 Event yield for different background processes, total expected background, and data in the signal region (SR). Final four rows show results in SR with sliding E_T^{miss} requirement

	Diboson	$W(\to \ell\nu)$ & $Z(\to \ell\ell)$ + jets	Top	Multijet	$Z(\to \nu\nu)$ + jets	$Vh(\to b\bar{b})$	Expected Backgrounds	Data
$\Delta\phi_{min}(j, E_T^{miss}) > 1$	25,496 ± 64	631,633 ± 565	29,573 ± 44	52,323 ± 3195	646,944 ± 2904	78 ± 2	1,386,046 ± 4355	1,385,095
$p_T(j_0) > 100$ GeV	25,450 ± 64	631,318 ± 565	29,524 ± 44	52,323 ± 3195	646,470 ± 2903	62 ± 1	1,385,146 ± 4354	1,383,975
At least one b-btag 60% eff.	1993 ± 17	16,858 ± 80	19,296 ± 33	7329 ± 2220	28,779 ± 683	44 ± 1	74,300 ± 2325	72,747
$(n_j \geq 3)$ $p_T(j_1) > 100$ GeV	1555 ± 15	13,131 ± 76	7459 ± 22	3439 ± 569	23,825 ± 617	33 ± 1	49,443 ± 843	51,174
$(n_j \geq 3)$ $p_T(b_1) > 60$ GeV	1550 ± 15	13,081 ± 76	7120 ± 21	3410 ± 568	23,752 ± 615	33 ± 1	48,946 ± 841	50,696
$p_t(b_0) > 100$ GeV	1164 ± 13	8606 ± 68	4378 ± 16	2589 ± 549	19,382 ± 558	26 ± 1	36,146 ± 787	35,215
$2 \leq n_j \leq 3$	673 ± 10	3865 ± 32	3079 ± 13	1291 ± 454	7781 ± 367	20 ± 1	16,708 ± 585	15,973
$\Delta R(j_0, j_1) < 2.0$	512 ± 9	1571 ± 18	1608 ± 9	353 ± 170	3675 ± 247	17 ± 1	7735 ± 301	7133
At least two b-btag 60% eff.	98 ± 4	114 ± 3	349 ± 4	28 ± 7	520 ± 117	8 ± 0	1115 ± 117	1027
90 GeV $\leq m_{bb} \leq$ 150 GeV	45 ± 2	35 ± 2	179 ± 3	11 ± 5	199 ± 79	7 ± 0	477 ± 79	467
$\Delta R(j_0, j_1) < 1.5$	37 ± 2	20 ± 1	83 ± 2	10 ± 5	94 ± 44	5 ± 0	250 ± 45	244
$E_T^{miss} \geq 150$ GeV	29.44 ± 1.93	14.55 ± 0.80	47.52 ± 1.55	3.65 ± 3.13	47.74 ± 25.85	4.95 ± 0.30	147.84 ± 26.17	164
$E_T^{miss} \geq 200$ GeV	13.17 ± 1.26	6.16 ± 0.51	17.18 ± 0.94	0.02 ± 0.02	21.18 ± 2.49	4.17 ± 0.25	61.88 ± 2.99	68
$E_T^{miss} \geq 300$ GeV	2.75 ± 0.58	1.13 ± 0.22	1.63 ± 0.27	0.00 ± 0.00	2.87 ± 0.96	1.04 ± 0.11	9.42 ± 1.18	11
$E_T^{miss} \geq 400$ GeV	0.59 ± 0.26	0.26 ± 0.10	0.28 ± 0.11	0.00 ± 0.00	0.33 ± 0.33	0.27 ± 0.05	1.73 ± 0.45	2

The uncertainties show statistical errors

Table 12.2 Exclusive (current yield divided by last yield) and inclusive (current yield divided by initial yield) efficiencies in the signal region for different background processes and total expected background

	Diboson		$W/Z+$ jets		$t\bar{t}$		Multijet		$Z \to \nu\nu$		$Vh(\to b\bar{b})$		Exp. Bkgd.	
	Excl.	Incl.	Excl.	Incl.	Excl.	Incl.	Excl.	Incl.	Excl.	Incl.	Excl.	Incl.	Excl.	Incl.
$1 \leq n_j \leq 4$	0.96	0.9603	0.95	0.9540	0.55	0.5501	0.76	0.7595	0.98	0.9779	0.92	0.9180	0.83	0.8278
$\Delta\phi_{\min}(j, E_T^{\text{miss}}) > 1$	0.74	0.7134	0.68	0.6492	0.43	0.2352	0.02	0.0170	0.84	0.8254	0.64	0.5842	0.34	0.2776
$p_T(j_0) > 100\,\text{GeV}$	1.00	0.7121	1.00	0.6489	1.00	0.2348	1.00	0.0170	1.00	0.8248	0.79	0.4640	1.00	0.2774
At least one b-btag 60 % eff.	0.08	0.0558	0.03	0.0173	0.65	0.1535	0.14	0.0024	0.04	0.0367	0.72	0.3326	0.05	0.0149
$(n_j \geq 3)\, p_T(j_1) > 100\,\text{GeV}$	0.78	0.0435	0.78	0.0135	0.39	0.0593	0.47	0.0011	0.83	0.0304	0.76	0.2513	0.67	0.0099
$(n_j \geq 3)\, p_T(b_1) > 60\,\text{GeV}$	1.00	0.0434	1.00	0.0134	0.95	0.0566	0.99	0.0011	1.00	0.0303	0.99	0.2484	0.99	0.0098
$p_t(b_0) > 100\,\text{GeV}$	0.75	0.0326	0.66	0.0088	0.61	0.0348	0.76	0.0008	0.82	0.0247	0.80	0.1988	0.74	0.0072
$2 \leq n_j \leq 3$	0.58	0.0188	0.45	0.0040	0.70	0.0245	0.50	0.0004	0.40	0.0099	0.74	0.1467	0.46	0.0033
$\Delta R(j_0, j_1) < 2.0$	0.76	0.0143	0.41	0.0016	0.52	0.0128	0.27	0.0001	0.47	0.0047	0.85	0.1253	0.46	0.0015
At least two b-btag 60 % eff.	0.19	0.0027	0.07	0.0001	0.22	0.0028	0.08	0.0000	0.14	0.0007	0.45	0.0567	0.14	0.0002
$90\,\text{GeV} \leq m_{bb} \leq 150\,\text{GeV}$	0.46	0.0013	0.31	0.0000	0.51	0.0014	0.41	0.0000	0.38	0.0003	0.92	0.0521	0.43	0.0001
$\Delta R(j_0, j_1) < 1.5$	0.83	0.0010	0.56	0.0000	0.47	0.0007	0.88	0.0000	0.47	0.0001	0.78	0.0404	0.53	0.0001
$E_T^{\text{miss}} \geq 150\,\text{GeV}$	0.79	0.0008	0.73	0.0000	0.57	0.0004	0.37	0.0000	0.51	0.0001	0.92	0.0372	0.59	0.0000
$E_T^{\text{miss}} \geq 200\,\text{GeV}$	0.45	0.0004	0.42	0.0000	0.36	0.0001	0.00	0.0000	0.44	0.0000	0.84	0.0314	0.42	0.0000
$E_T^{\text{miss}} \geq 300\,\text{GeV}$	0.21	0.0001	0.18	0.0001	0.10	0.0000	0.00	0.0000	0.14	0.0000	0.25	0.0078	0.15	0.0000
$E_T^{\text{miss}} \geq 400\,\text{GeV}$	0.22	0.0000	0.23	0.0000	0.17	0.0000	1.00	0.0000	0.12	0.0000	0.26	0.0021	0.19	0.0000

12.1 Dark Matter + Higgs($\to b\bar{b}$) Signal Region

Table 12.3 The numbers of predicted background events for each background process, the sum of all background components, and observed data in the signal region (SR) of the resolved and boosted channels for each of the sliding E_T^{miss} requirements

E_T^{miss}	Resolved				Boosted	
	> 150 GeV	> 200 GeV	> 300 GeV	> 400 GeV	> 300 GeV	> 400 GeV
$Z(\to \nu\bar{\nu})$+jets	48 ± 32	21 ± 5	2.9 ± 1.1	0.3 ± 0.3	7.0 ± 2.0	5.2 ± 1.6
Multijet	3.7 ± 3.1	0.02 ± 0.02	–	–	< 0.0 ± 0.1	< 0.0 ± 0.1
$t\bar{t}$ & single-top	48 ± 10	17 ± 3.8	1.6 ± 0.5	0.3 ± 0.1	0.8 ± 0.5	0.6 ± 0.4
W/Z+jets	15 ± 3.4	6.2 ± 1.5	1.1 ± 0.3	0.3 ± 0.1	1.4 ± 0.7	0.8 ± 0.4
Diboson	29.4 ± 7.5	13.2 ± 3.8	2.8 ± 1.0	0.6 ± 0.3	0.9 ± 0.5	0.6 ± 0.3
$Vh(\to b\bar{b})$	5.0 ± 0.7	4.2 ± 0.6	1.0 ± 0.2	0.3 ± 0.1	1.0 ± 0.2	0.6 ± 0.1
Total Bkgd.	148 ± 30	62 ± 7.5	9.4 ± 1.8	1.7 ± 0.5	11.2 ± 2.3	7.7 ± 1.7
Data	164	68	11	2	20	9

Statistical and systematic uncertainties are combined. The uncertainties on the total background take into account the correlation of systematic uncertainties among different background processes. The large uncertainty on the $Z(\to \nu\bar{\nu})$+jets process in the $E_T^{miss} > 150$ GeV SR of the resolved channel is due to limited statistics in the $Z(\to \mu^+\mu^-)$+jets data sample used for the estimation of $Z(\to \nu\bar{\nu})$+jets with $E_T^{miss} < 200$ GeV

12.1.2 Final Selections for DM + $h(b\bar{b})$ by $m_{Z'}$ and m_A

Based on the event yield and selection efficiencies for the simulated $Z' \to Ah$ signal process, and the total expected background as detailed in the previous section, we finalize the sliding E_T^{miss} requirement in the ($m_{Z'}, m_A$) parameter space to maximize signal sensitivity. The results are given in Table 12.4, with the expected signal yield for the chosen sliding E_T^{miss} requirement at the specific ($m_{Z'}, m_A$) in boldface. The expected signal yield values are for $\tan\beta = 1$ and scaled to the maximum allowed g_z value at the respective $m_{Z'}$ as shown in Fig. 7.2 and Table 7.1. The projected signal sensitivity is calculated as:

$$S = \frac{N_S}{\sqrt{N_S + N_b + 0.2^2 \cdot N_b^2}} \quad (12.1)$$

N_S refers to the number of expected signal events. N_b refers to the total expected background yield for said final cut on E_T^{miss}. The third term in the denominator is a rough approximation of systematic errors which were conservatively estimated to be 20% of the number of background events.

The event yields at each of the E_T^{miss} requirements from the $Z' \to Zh$ process that also contributes to the $E_T^{miss} + h(\to b\bar{b})$ final state are given in Table 12.5. The expected signal yield values are for $\tan\beta = 1$ and scaled to the maximum allowed g_z value at the respective $m_{Z'}$ as shown in Fig. 7.2 and Table 7.1. In the final limit setting process, in addition to results for $Z' \to Ah$ process alone, the $Z' \to Zh$ contribution is added to the $Z' \to Ah$ yield, and limits are set both for selections

Table 12.4 Expected signal and background yield with sliding E_T^{miss} requirement to select DM + $h(b\bar{b})$ signal from the $Z' \to Ah$ process with varying $m_{Z'}$ and m_A

$m_{Z'}$	m_A	$E_T^{miss} \geq 150$	$E_T^{miss} \geq 200$	$E_T^{miss} \geq 300$	$E_T^{miss} \geq 400$	S
600	300	**10.3**	3.5	0.13	0.02	0.36
600	400	**0.3**	0.14	0.03	0.01	0.011
800	300	23.6	22.6	**10.4**	0.24	2.36
800	400	6.8	**6.0**	0.90	0.03	0.46
800	500	2.0	**1.24**	0.08	0.01	0.10
800	600	0.26	**0.15**	0.03	0.01	0.012
1000	300	14.3	14.1	12.5	**6.4**	2.35
1000	400	5.8	5.7	4.8	**1.3**	0.86
1000	500	3.3	3.2	**2.14**	0.19	0.64
1000	600	1.64	1.52	**0.51**	0.04	0.17
1000	700	0.56	**0.42**	0.09	0.01	0.033
1000	800	0.13	**0.09**	0.03	0.01	0.01
1200	300	9.6	9.6	9.3	**7.8**	2.62
1200	400	4.5	4.5	4.2	**3.3**	1.58
1200	500	3.1	3.0	2.8	**2.0**	1.15
1200	600	2.2	2.1	1.9	**0.93**	0.63
1200	700	1.3	1.3	**1.0**	0.23	0.32
1200	800	0.7	0.6	**0.31**	0.06	0.10
1400	300	2.5	2.5	2.5	**2.2**	1.22
1400	400	1.3	1.3	1.3	**1.1**	0.75
1400	500	1.1	1.1	1.1	**0.94**	0.74
1400	600	0.94	0.94	0.90	**0.77**	0.57
1400	700	0.75	0.75	0.71	**0.54**	0.43
1400	800	0.51	0.50	0.45	**0.28**	0.24
Tot. Exp. Bkgd.		148	62	9.4	1.7	

Boldfaced numbers reflect the E_T^{miss} requirement with best sensitivity S [also shown, calculated per Eq. (12.1)] for the $Z' \to Ah$ process. Masses are in *GeV*

optimized for $Z' \to Ah$ alone as in Table 12.4, and for selections optimized for the two processes combined as shown in Table 12.6. As $Z' \to Zh$ makes a considerable contribution to regions of parameter space probed in this analysis, and from an experimental standpoint, the two processes both make up the $E_T^{miss} + h(\to b\bar{b})$ final state, the final results described in Sect. 12.3 are for both processes combined and the selection optimized for the total yield.

12.1.3 Kinematic Distributions

We compare the kinematic distributions between data and expected background in the signal region through each of the selection stages. Selected variables from

12.2 Statistical Interpretation

Table 12.5 Expected signal yield from $Z' \to Zh$ process in the final signal region with sliding E_T^{miss} requirement. Masses are in *GeV*

$m_{Z'}$	$E_T^{miss} \geq 150$	$E_T^{miss} \geq 200$	$E_T^{miss} \geq 300$	$E_T^{miss} \geq 400$
600	13.6	11.9	0.8	0.02
800	9.2	9.0	6.8	0.8
1000	4.2	4.1	3.8	2.7
1200	2.4	2.4	2.3	2.0
1400	0.5	0.5	0.5	0.4
Tot. Exp. Bkgd.	148	62	9.4	1.7

representative selection stages are included here to show good modeling between data and background, and that no significant excess is observed. Figure 12.1 shows the comparison of data and expected background for $2 \leq n_j \leq 3$, at least two b-jet in the full range of E_T^{miss} above 100 GeV. For the above sets of plots, to preserve statistics, we choose the full E_T^{miss} range above 100 GeV, and ease the cut on $\Delta R(p_T(j0), p_T(j1))$ to below 2.0 instead of 1.5 as is in the final signal selection.

Figure 12.2 shows the comparison of data, expected background, and select signal samples, in the final signal region for E_T^{miss} above 150 GeV. The signal distributions are not scaled to cross-section and show shape only.

As the standard model $Vh(\to b\bar{b})$ production is small compared with the rest of the background, and is partially included in the $Z \to \nu\bar{\nu}$ data-driven estimation for E_T^{miss} below 200 GeV as explained earlier, it is not plotted alone: instead, for Fig. 12.2, the $Z(\to \nu\nu) + h(\to b\bar{b})$ contributions (with E_T^{miss} above 200 GeV) are added to the $Z \to \nu\bar{\nu}$ background, and the $W(\to \ell\nu) + h(\to b\bar{b})$ process (in the full E_T^{miss} range) is included in the $W(\to \ell\nu)/Z(\to \ell\ell)$+jets background.

12.2 Statistical Interpretation

The data sample obtained after full selection is compared to the expected sample from background processes (both data-driven and MC) and the statistical significance of a potential excess or the limit on signal production is estimated. We use the program `Histfitter 00-00-47` to estimate an upper limit on the maximal signal strength μ_{sig} of our signal samples. The `Histfitter` package uses a technique called profile maximization (profiling) which refers to the practice of determining the "best fit" of the predicted model to data by maximizing the likelihood over the possible values of nuisance parameters.[1]

Given a set of predictions, observations, and systematic uncertainties, one can define a model which represents the best fit to the data observation within the

---

[1] In statistics, a nuisance parameter is any parameter which is not of immediate interest but which must be accounted for in the analysis of those parameters which are of interest.

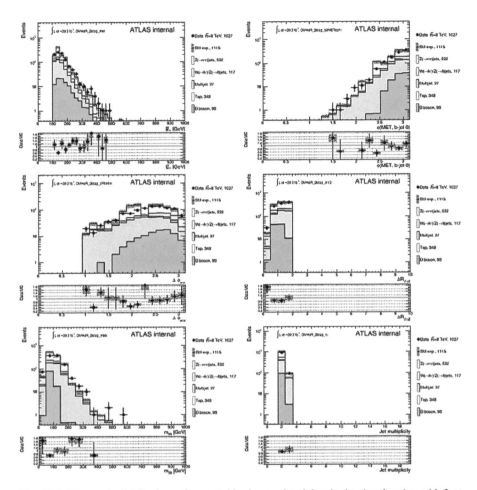

Fig. 12.1 Kinematic distributions of expected background and data in the signal region with $2 \leq n_j \leq 3$, at least two b-jets in the full range of E_T^{miss} above 100 GeV

constraints of the systematic uncertainties. A reliable method of performing this fit is achieved by maximizing the likelihood function. The inputs to the function are the best estimates for each background source and nuisance parameter. The general likelihood L of the analyses considered here is the product of Poisson distributions of event counts in the SR and of additional distributions that implement the constraints on systematic uncertainties. It can be written as:

$$L(\boldsymbol{n}, \boldsymbol{\theta}^0 | \mu_{\text{sig}}, \boldsymbol{b}, \boldsymbol{\theta}) = P(n_S | \lambda_S(\mu_{\text{sig}}, \boldsymbol{b}, \boldsymbol{\theta})) \times C_{\text{syst}}(\boldsymbol{\theta}^0, \boldsymbol{\theta}) \quad (12.2)$$

The first factor of Eq. (12.2) reflects the Poisson measurements of n_S, the number of observed events in the signal region. The Poisson expectation λ_S depends on

12.2 Statistical Interpretation

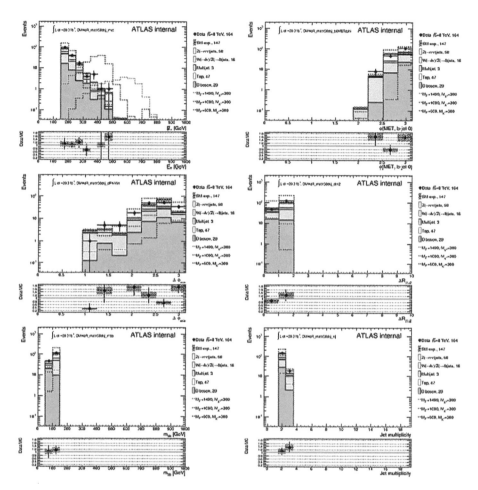

Fig. 12.2 Kinematic distributions of expected background and select signal samples in final signal region with $E_\mathrm{T}^\mathrm{miss}$ above 150 GeV. The $E_\mathrm{T}^\mathrm{miss}$ distributions for a few signal processes are overlayed in *dashed lines* for shape comparison. The small contribution from Zh (Wh) process is included in the $Z(\to \nu\bar{\nu})$+jets ($W(\to \ell\nu)/Z(\to \ell\ell)$+jets) distribution

the predictions **b** for various background sources, the nuisance parameters that parametrize systematic uncertainties, and also the signal strength parameter μ_sig. For $\mu_\mathrm{sig} = 0$ the signal component is turned off for a "background-only" likelihood, and for $\mu_\mathrm{sig} = 1$ the signal expectation equals the nominal value of the model under consideration. $C_\mathrm{syst}(\boldsymbol{\theta}^0, \boldsymbol{\theta})$ is the probability density function that implements the constraints on systematic uncertainties, where $\boldsymbol{\theta}$ represents the systematic uncertainties, and θ^0 the central value around which $\boldsymbol{\theta}$ may be varied.

By LHC default, Histfitter employs a Frequentist method to perform hypothesis tests and uses the profile likelihood ratio $q_{\mu_{\text{sig}}}$ as test statistic [1], defined as

$$q_\mu = -2\log(L(\mu, \hat{\hat{\theta}})/L(\hat{\mu}, \hat{\theta})) \qquad (12.3)$$

where the denominator is the "likelihood of best fit," and the numerator is the "likelihood assuming μ signal strength.".

This approach ensures that uncertainties and their correlations are propagated to the outcome with their proper weights.

In case of discovery, the "Null" hypothesis q_0 describes the background only, while the alternative "Test" hypothesis describes signal plus background. In the classical approach, one first requires that the p-value of q_0 is found below the given threshold (in HEP one requires $p \leq 2.87 \times 10^{-7}$). If this condition is satisfied, one looks for an alternative hypothesis which can explain the data well. Considering there are four sliding E_T^{miss} windows for the signal region, the "look elsewhere" approach will be adopted in calculating the p-value.

12.3 Constraints on Z'-2HDM Model

In this analysis, we adopt a simple "cut and count" method, where the limit setting does not depend on the shape of distributions. The limits in this analysis are derived using the CL_s method with a log-likelihood ratio (LLR) test statistic. For the systematics uncertainties listed in Tables 11.7, 11.8, 11.9, and 11.10, the experimental uncertainties on the signal and background were considered fully correlated, while the PDF and cross-section uncertainties, i.e., the theoretical uncertainties, are treated as uncorrelated. For each signal sample, in addition to the aforementioned systematic uncertainties, the number of observed events, expected background, and the signal, as well as statistical uncertainties on the last two, are used as input for the calculation. To set the upper limit, a set of exclusion hypothesis tests are performed using the asymptotic calculator and one-sided profile likelihood test statistics, varying the signal strength and interpolating for the signal strength where 95 % exclusion is achieved. An example of the upper limit plot is shown in Fig. 12.3.

While the mono-Higgs signal associated with dark matter production comes from the $Z' \to hA$, $h \to bb$, $A \to \chi\chi$ process, as was mentioned in Sect. 7.1, the Z'-2HDM model used for this analysis enjoys an additional source of Higgs plus E_T^{miss} from the decay of $Z' \to hZ$, where Z boson decays to a pair of neutrinos, i.e., invisibly. The two processes, as shown in Fig. 7.3, have very similar kinematics with a slightly harder E_T^{miss} spectrum for $Z' \to Zh$ due to the smaller Z mass compared to m_A (≥ 300 GeV).

Considering the considerable contribution from $Z' \to Zh$ to the $h(\to b\bar{b})$ + E_T^{miss} final state, we reoptimize the selection taking into account the combined signal

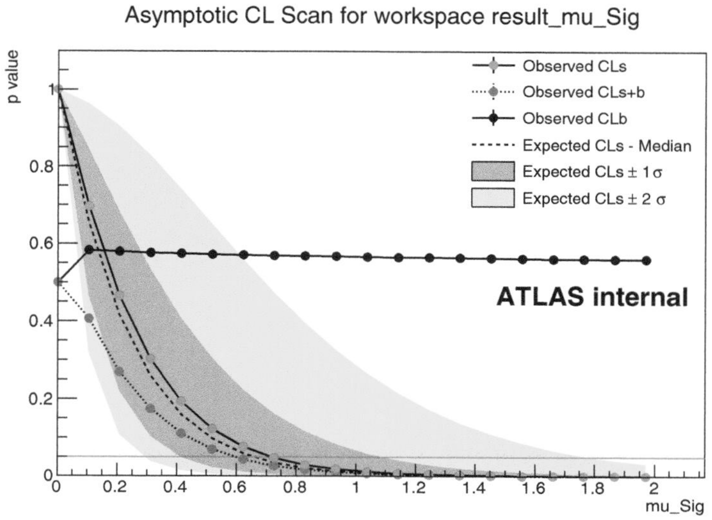

12.3 Constraints on Z'-2HDM Model

Fig. 12.3 Upper limit for $Z' \to Ah$ signal with $m_{Z'} = 1000\,\text{GeV}$, $m_A = 300\,\text{GeV}$

yield from both processes, and calculate the limits in both the $m_{Z'}$–m_A plane and $m_{Z'}$–$\tan\beta$ plane described earlier. Depending on the specific parameter space, either $Z' \to Ah$ or $Z' \to Zh$ may constitute the majority contribution to the final $E_T^{\text{miss}} + h(\to b\bar{b})$ signal. For both cases, the Z' gauge coupling is set to its 95% CL upper limit from precision electroweak constraints and searches for dijet resonances for the corresponding Z' mass and $\tan\beta$ value. The Z' boson does not couple to leptons in this model, avoiding potentially stringent constraints from dilepton searches. Taking the alignment limit of $\alpha = \beta - \pi/2$ evades the constraints in $\tan\beta$ for a type 2 two-Higgs-doublet model using fits to the observed Higgs boson couplings from the LHC [2].

It is worth noting that for this analysis, the limit calculation is conservative, as the signal cross-sections are calculated at tree-level with K-factor set to 1. Adding QCD corrections to the signals, including ISR corrections, could increase the signal cross-section, hence placing a stronger limit on the signals.

$m_{Z'}$–m_A Plane For the m'_Z–m_A plane, Table 12.6 shows the modified E_T^{miss} selection optimized for $Z' \to Ah$ and $Z' \to hZ$ combined signal, the total and individual signal contributions from either process, and the observed and computed upper limits on signal strength. The exclusion region in the $m_{Z'}$–m_A plane is shown in Fig. 12.4, where $m_A \geq 300\,\text{GeV}$ in accordance with $b \to s\gamma$ constraints [3].

For $\tan\beta = 1$, where theoretical projections indicate sensitivity in a search for Z'-2HDM signal production in the $b\bar{b}$ channel at $\sqrt{s} = 8\,\text{TeV}$ [4], $m_{Z'} = 700$–$1300\,\text{GeV}$ is excluded for m_A up to 350 GeV, with further exclusion of larger m_A for $m_{Z'}$ around 1200 GeV. In general, better sensitivity is reached at larger $m_{Z'}$ and lower m_A, partly because it yields a harder E_T^{miss} spectrum, allowing us to

Table 12.6 $Z' \to Ah$ plus $Z' \to hZ$ signal sensitivity, varying $m_{Z'}$ and m_A (units in GeV, $m_A \geq 300$ GeV in accordance with $b \to s\gamma$ constraints) at $\tan\beta = 1$ with a sliding E_T^{miss} cut (in GeV) applied to optimize sensitivity for both $Z' \to Ah$ and $Z' \to hZ$ processes combined

$m_{Z'}$	m_A	E_T^{miss}	N_{obs}	N_{bkgd}	N_{sig}	N_{sig}^{Ah}	N_{sig}^{Zh}	μ_{sig}	Lim$_{\text{Obs}}$	Lim$_{\text{Exp.}}$	Lim$_{\text{Exp.}}^{+1\sigma}$	Lim$_{\text{Exp.}}^{-1\sigma}$
600	300	≥ 200	68	62	15.4	3.5	11.9	0.41	2.00	1.62	2.43	1.12
600	400	≥ 200	68	62	12.0	0.1	11.9	0.51	2.57	2.08	3.12	1.43
800	300	≥ 300	11	9.4	17.2	10.4	6.8	0.10	0.58	0.49	0.75	0.33
800	400	≥ 300	11	9.4	7.7	0.9	6.8	0.22	1.34	1.12	1.74	0.75
800	500	≥ 300	11	9.4	6.9	0.1	6.8	0.24	1.50	1.25	1.94	0.84
800	600	≥ 300	11	9.4	6.8	0.0	6.8	0.25	1.53	1.27	1.97	0.85
1000	300	≥ 400	2	1.7	9.1	6.4	2.7	0.03	0.50	0.46	0.76	0.29
1000	400	≥ 400	2	1.7	4.0	1.3	2.7	0.08	1.16	1.07	1.77	0.67
1000	500	≥ 400	2	1.7	2.9	0.2	2.7	0.11	1.62	1.49	2.48	0.94
1000	600	≥ 400	2	1.7	2.7	0.0	2.7	0.11	1.74	1.60	2.66	1.01
1000	700	≥ 400	2	1.7	2.7	0.0	2.7	0.12	1.82	1.67	2.78	1.05
1000	800	≥ 400	2	1.7	2.7	0.0	2.7	0.12	1.89	1.74	2.89	1.10
1200	300	≥ 400	2	1.7	9.8	7.8	2.0	0.03	0.47	0.43	0.71	0.27
1200	400	≥ 400	2	1.7	5.3	3.3	2.0	0.07	0.87	0.81	1.33	0.51
1200	500	≥ 400	2	1.7	4.0	2.0	2.0	0.08	1.16	1.07	1.77	0.67
1200	600	≥ 400	2	1.7	3.0	1.0	2.0	0.10	1.57	1.44	2.39	0.91
1200	700	≥ 400	2	1.7	2.3	0.2	2.0	0.13	2.07	1.89	3.16	1.19
1200	800	≥ 400	2	1.7	2.1	0.1	2.0	0.15	2.27	2.08	3.48	1.31
1400	300	≥ 400	2	1.7	2.6	2.2	0.4	0.12	1.81	1.66	2.77	1.05
1400	400	≥ 400	2	1.7	1.5	1.1	0.4	0.20	3.21	2.91	4.93	1.85
1400	500	≥ 400	2	1.7	1.4	1.0	0.4	0.22	3.50	3.17	5.38	2.02
1400	600	≥ 400	2	1.7	1.2	0.8	0.4	0.30	4.12	3.75	6.36	2.38
1400	700	≥ 400	2	1.7	1.0	0.6	0.4	0.35	4.99	4.55	7.75	2.88
1400	800	≥ 400	2	1.7	0.7	0.3	0.4	0.46	7.26	6.68	11.49	4.20

The columns from left to right describe $m_{Z'}$, m_A, the final sliding E_T^{miss} cut used, the observed data in the SR, the expected background, the expected signal yield, the calculated signal strength, the observed and expected (median, $+1\sigma$, and -1σ) 95% CL upper limit on signal strength. The stronger limit reflects the additional $h(\to b\bar{b}) + E_T^{\text{miss}}$ contribution from the $Z' \to hZ$ process

adopt a higher E_T^{miss} requirement for better background suppression; there is also a larger Z' gauge coupling allowed at higher $m_{Z'}$, but for very large $m_{Z'}$, the signal production cross-section eventually decreases due to PDF suppression, leading to lower sensitivity.

The decay width for the $Z' \to Ah$ signal process is given in Eq. (7.8), while the decay width for the $Z' \to Zh$ process is given in Eq. (7.9). As α is taken to be $\beta - \pi/2$ for the alignment limit, the production cross-section of the two processes become comparable at $\tan\beta = 1$. For $m_A = 300$ GeV, $Z' \to hZ$ contributes about a quarter of the total $h(\to b\bar{b}) + E_T^{\text{miss}}$ signal, largely corresponding to the difference

12.3 Constraints on Z′-2HDM Model

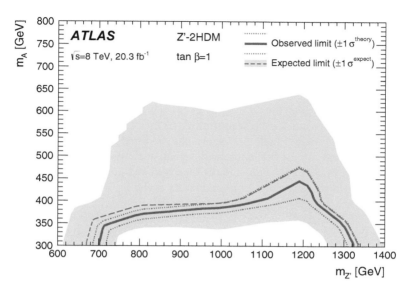

Fig. 12.4 Exclusion contour for $Z' \to Ah$ plus $Z' \to hZ$ combined in the $m_{Z'}$–$\tan\beta$ plane, with a sliding E_T^{miss} requirement (in GeV) applied to optimize sensitivity for both $Z' \to Ah$ and $Z' \to hZ$ processes combined ($\tan\beta \geq 0.3$ based on perturbativity of the top Yukawa coupling and $m_A \geq 300\,\text{GeV}$). The expected limit is given by the *dashed blue line*, and the *yellow bands* indicate its $\pm 1\sigma$ uncertainty. The observed limit is given by the *solid red line*, and the *red dotted lines* show the variations of the observed limit due to a $\pm 1\sigma$ change in the signal theoretical cross-section. The parameter spaces below the limit contours are excluded at 95 % CL

in branching ratio of $A \to \chi\chi$, which is close to 100 %, compared to that of $Z \to \nu\nu$. For $m_A \geq 400\,\text{GeV}$, $Z' \to hZ$ contributes a considerably larger fraction of the final signal, partly due to the additional decay mode of $A \to t\bar{t}$ leading to a smaller dark matter production cross-section, and partly due to the softer E_T^{miss} spectrum for the $Z' \to Ah$ process when A is much heavier than the Z'.

12.3.1 $m_{Z'}$–$\tan\beta$ Plane

While the signal samples in this analysis are produced at a fixed $\tan\beta$ value of 1 and $g_z = 0.8$, the $Z' \to Ah$ cross-section can be scaled as a function of $\tan\beta$ as in Eq. (7.8), where $\alpha = \beta - \pi/2$, and the upper limit on g_z is given by ρ_0 constraints and dijet measurements (Fig. 7.2). Hence, we take $m_A = 300\,\text{GeV}$ which yields the largest dark matter signal production, varying m_Z' in the aforementioned space of 600–1400 GeV, and calculate the expected signal yield varying values of $\tan\beta$, with $\tan\beta \geq 0.3$ based on the perturbativity requirement of the Higgs–top Yukawa coupling [5].

When $\tan\beta \lesssim 0.6$, the constraint on g_z is independent of $\tan\beta$ as the dijet constraints are stronger (Fig. 7.2), and the width is therefore suppressed by $\sin^4\beta$; hence at small $\tan\beta$, $Z' \to Ah$ is the dominant process.

At large $\tan\beta$, $Z' \to Ah$ is suppressed by $(\sin\beta\cos\beta)^2$ (Eq. (7.8), where $\alpha = \beta - \pi/2$), and $Z' \to hZ$ becomes the dominant process. Although the $Z' \to hZ$ rate naively scales as $\sin^4\beta$ from Eq. (7.9), this dependence is almost exactly cancelled when we apply the upper limit on g_z from ρ_0, which leads to an upper limit on $g_z \propto 1/(\sin^2\beta)$. This can also be seen from Eqs. (7.4), (7.5).

For the m'_Z–$\tan\beta$ plane, Table 12.7 shows the modified E_T^{miss} selection optimized for $Z' \to Ah$ and $Z' \to hZ$ combined signal, the total and individual signal contributions from either process, and the observed and computed upper limits on signal strength. Two-dimensional limit contours in the $m_{Z'}$–$\tan\beta$ plane are shown in Fig. 12.5, where $\tan\beta$ is ≥ 0.3 based on the perturbativity requirement of the Higgs–top Yukawa coupling, and is below 10 based on direct searches for the A [6]. For $m_A = 300\,\mathrm{GeV}$, where A decays almost exclusively to a DM pair, $m_{Z'} = 700$–$1300\,\mathrm{GeV}$ is excluded for $\tan\beta < 2$, with further exclusion of larger $\tan\beta$ for $m_{Z'}$ between 800 and 1000 GeV due to the inclusion of the $Z' \to Zh$ contribution in the final state.

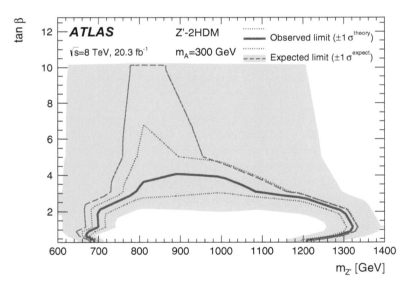

Fig. 12.5 Exclusion contour for $Z' \to Ah$ plus $Z' \to hZ$ combined in the $m_{Z'}$–$\tan\beta$ plane, with a sliding E_T^{miss} requirement (in *GeV*) applied to optimize sensitivity for both $Z' \to Ah$ and $Z' \to hZ$ processes combined ($\tan\beta \geq 0.3$ based on perturbativity of the top Yukawa coupling and $m_A = 300\,\mathrm{GeV}$). The expected limit is given by the *dashed blue line*, and the *yellow bands* indicate its $\pm 1\sigma$ uncertainty. The observed limit is given by the *solid red line*, and the *red dotted lines* show the variations of the observed limit due to a $\pm 1\sigma$ change in the signal theoretical cross-section. The parameter spaces below the limit contours are excluded at 95 % CL

12.3 Constraints on Z'-2HDM Model

Table 12.7 $Z' \to Ah$ plus $Z' \to hZ$ combined signal sensitivity, varying $m_{Z'}$ (units in GeV) and $\tan\beta$ (≥ 0.3 based on perturbativity of the top Yukawa coupling) at $m_A = 300\,\text{GeV}$ with a sliding E_T^{miss} cut (in GeV) applied to optimize sensitivity for both $Z' \to Ah$ and $Z' \to hZ$ processes combined

$m_{Z'}$	$\tan\beta$	E_T^{miss}	N_{obs}	N_{bkgd}	N_{sig}	N_{sig}^{Ah}	N_{sig}^{Zh}	μ_{sig}	Lim_{Obs}	$\text{Lim}_{\text{Exp.}}$	$\text{Lim}_{\text{Exp.}}^{+1\sigma}$	$\text{Lim}_{\text{Exp.}}^{-1\sigma}$
600	0.3	≥ 150	164	148	11.1	10	1.1	1.45	9.28	7.33	12.08	4.79
600	0.5	≥ 150	164	148	28.3	21.3	7	0.58	3.33	2.65	4.25	1.76
600	1	≥ 200	68	62	15.4	3.5	11.9	0.41	2.00	1.62	2.43	1.12
600	3	≥ 200	68	62	16.3	0.5	15.8	0.39	1.89	1.53	2.29	1.06
600	5	≥ 200	68	62	16.0	0.2	15.8	0.40	1.93	1.56	2.34	1.08
600	10	≥ 200	68	62	15.8	0.0	15.8	0.40	1.95	1.58	2.37	1.09
800	0.3	≥ 300	11	9.4	13.2	12.5	0.7	0.12	0.77	0.64	1.00	0.43
800	0.5	≥ 300	11	9.4	31.0	26.6	4.4	0.05	0.32	0.27	0.41	0.18
800	1	≥ 300	11	9.4	17.2	10.4	6.8	0.10	0.58	0.49	0.75	0.33
800	3	≥ 300	11	9.4	10.5	1.5	9.0	0.15	0.98	0.81	1.26	0.55
800	5	≥ 300	11	9.4	9.5	0.5	9.0	0.17	1.08	0.90	1.40	0.61
800	10	≥ 300	11	9.4	9.0	0.0	9.0	0.18	1.14	0.95	1.48	0.64
1000	0.3	≥ 400	2	1.7	12.6	12.2	0.4	0.03	0.36	0.34	0.55	0.21
1000	0.5	≥ 400	2	1.7	26.0	25.6	0.4	0.01	0.18	0.16	0.27	0.10
1000	1	≥ 400	2	1.7	9.1	6.4	2.7	0.03	0.50	0.46	0.76	0.29
1000	3	≥ 400	2	1.7	5.1	1.1	4.0	0.07	0.91	0.84	1.38	0.53
1000	5	≥ 400	2	1.7	4.2	0.4	3.9	0.08	1.13	1.05	1.73	0.66
1000	10	≥ 400	2	1.7	3.9	0.0	3.9	0.09	1.26	1.16	1.92	0.73
1200	0.3	≥ 400	2	1.7	2.9	2.8	0.1	0.11	1.62	1.49	2.48	0.94
1200	0.5	≥ 400	2	1.7	6.4	6.0	0.4	0.05	0.72	0.67	1.10	0.42
1200	1	≥ 400	2	1.7	9.8	7.8	2	0.03	0.47	0.43	0.71	0.27
1200	3	≥ 400	2	1.7	4.1	1.3	2.8	0.08	1.14	1.05	1.73	0.66
1200	5	≥ 400	2	1.7	3.3	0.5	2.8	0.10	1.42	1.31	2.16	0.82
1200	10	≥ 400	2	1.7	2.8	0.0	2.8	0.11	1.68	1.55	2.57	0.97
1400	0.3	≥ 400	2	1.7	0.7	0.7	0.0	0.48	7.59	6.99	12.05	4.39
1400	0.5	≥ 400	2	1.7	1.5	1.4	0.1	0.21	3.25	2.95	5.00	1.88
1400	1	≥ 400	2	1.7	2.6	2.2	0.4	0.12	1.82	1.67	2.79	1.05
1400	3	≥ 400	2	1.7	1.6	0.6	1.0	0.19	3.04	2.76	4.66	1.75
1400	5	≥ 400	2	1.7	1.2	0.2	1.0	0.30	4.11	3.74	6.35	2.38
1400	10	≥ 400	2	1.7	1.0	0.0	1.0	0.35	4.99	4.55	7.75	2.88

The columns from left to right describe $m_{Z'}$, $\tan\beta$, the final sliding E_T^{miss} cut used, the observed data in the SR, the expected background, the expected signal yield, the calculated signal strength, the observed and expected (median, $+1\sigma$, and -1σ) 95% CL upper limit on signal strength

We compare Fig. 12.5 with the theoretical prediction of total cross-section of h + E_T^{miss} in $m_{Z'}$–tan β plane (Fig. 5. of [4]). Using the upper limit on visible cross-section from our search (Table 12.8), and the signal acceptance for both processes at the aforementioned E_T^{miss} requirement for given $m_{Z'}$ (Table 9.2 for $Z' \to Ah$, and Table 9.8 for $Z' \to hZ$), our limits are consistent with the theoretical prediction. Sensitivity is limited at low $m_{Z'}$ due to the larger background associated with the lower E_T^{miss} requirement ($\geq$ 150 GeV for $m_{Z'}$ = 600 GeV). At larger $m_{Z'}$, we adopt E_T^{miss} requirements $\geq$ 300 GeV for $m_{Z'}$ = 800 GeV, and $\geq$ 400 GeV for $m_{Z'}$ = 1000 GeV and above, leading to minimal background and excellent sensitivity in these signal regions. As the cross-section of the $Z' \to hZ$ process is approximately constant for a given $m_{Z'}$ at large tan β where $Z' \to Ah$ is suppressed, including the $Z' \to hZ$ contribution gives us the opportunity to probe and set limits on regions with large tan β. While the sensitivity can extend to much larger tan β, we apply an upper limit of tan $\beta < 10$ for m_A = 300 GeV from direct searches of A [6] and [7]. The regions with large tan β are also of less interest in this search due to its minimal dark matter production with the suppressed $Z' \to Ah$ process. As the total cross-section falls with larger $m_{Z'}$ due to PDF suppression, we have the best sensitivity for $m_{Z'}$ between 800 GeV and 1 TeV in the parameter space we probe.

We achieve much better exclusion limits compared with Fig. 7. of Ref. [4], as the latter adopts signal selection and background estimation from the existing ATLAS $Vh(\to b\bar{b})$ analysis [8], which uses much softer E_T^{miss} requirements of $\geq$ 120, $\geq$ 160, and $\geq$ 200 GeV. We probe much higher E_T^{miss} regions with dedicated selections optimized for the $h(\to b\bar{b}) + E_T^{miss}$ signal, and the improved limits further demonstrate the strength and novelty of this analysis.

12.4 Model-Independent Upper Limit

In addition to constraints on the Z'-2HDM model, we also calculate model-independent upper limits on the number of non-SM events in the signal region. Again for completeness, results from both the resolved and boosted channels are presented in this section, though as discussed in Chap. 6, the boosted analysis is not used to interpret the Z'-2HDM model. For each sliding E_T^{miss} requirement in the final signal selection, the expected background, including its statistical and systematic uncertainties, is fit to the number of observed events in a "null" hypothesis.

$$q_0 = -2\log(L(0, \hat{\hat{\theta}})/L(\hat{\mu}, \hat{\theta})) \tag{12.4}$$

where the denominator is the "likelihood of best fit," and the numerator is the "likelihood assuming background only."

The probability of the background-only hypothesis, the $p(s = 0)$-value, is calculated for each of the four signal regions with ascending E_T^{miss} threshold in the resolved channel and the two signal regions in the boosted channel using 3000 toy

Table 12.8 Model-independent upper limits for the resolved and boosted channels

	E_T^{miss}	N_{obs}	N_{bkgd}	$\langle\sigma_{vis}\rangle_{obs}^{95}$ [fb]	$N_{BSM\,obs}^{95}$	$N_{BSM\,exp}^{95}$	$p(s=0)$
Resolved	> 150 GeV	164	148	3.6	74	63^{+22}_{-14}	0.31
	> 200 GeV	68	62	1.3	27	$21^{+8.4}_{-3.9}$	0.28
	> 300 GeV	11	9.4	0.49	9.9	$8.2^{+3.4}_{-1.9}$	0.31
	> 400 GeV	2	1.7	0.24	4.8	$4.7^{+1.6}_{-1.0}$	0.39
Boosted	> 300 GeV	20	11.2	0.90	18	$9.9^{+4.2}_{-2.9}$	0.03
	> 400 GeV	9	7.7	0.43	8.8	$7.7^{+3.3}_{-2.0}$	0.37

Left to right: signal region (SR) E_T^{miss} requirement, number of observed events, number of expected background events, 95% CL upper limits on the visible cross-section ($\langle\sigma_{vis}\rangle_{obs}^{95}$), and the number of non-SM events ($N_{BSM\,obs}^{95}$). The sixth column ($N_{BSM\,exp}^{95}$) shows the expected 95% CL upper limit on the number of non-SM events, given the estimated number and the $\pm 1\sigma$ uncertainty of background events. The last column shows the p-value for the background-only hypothesis ($p(s=0)$)

experiments. Table 12.8 gives the model-independent 95% CL upper limits on the visible cross-section, defined as the product of production cross-section, acceptance, and reconstruction efficiency, the observed and expected limits on the number of non-Standard Model events in the signal region, and the null-hypothesis p-value.

As a $p(s=0)$-value of 0.03 is calculated for $E_T^{miss} > 300$ GeV in the boosted channel, a calculation of the look-elsewhere effect [9] is performed. The signal regions in both the resolved and boosted channels are divided into exclusive regions of E_T^{miss}, taking into account the small overlap between both channels ($\sim$15%). The event yield in each of the exclusive regions is fluctuated independently through a large number of pseudo-experiments. The results are added back into the final signal regions with the sliding E_T^{miss} requirements, and the $p(s=0)$-values are calculated for each of the variations. The trials factor for the look-elsewhere effect is calculated to be $\sim$3 for all regions combined using 10,000 pseudo-experiments, indicating there is a $\sim$10% likelihood that the small excess observed in this region is due to statistical fluctuations in the background.

References

1. M. Baak, G.J. Besjes, D. Côte, A. Koutsman, J. Lorenz et al., HistFitter software framework for statistical data analysis. Eur. Phys. J. **C75**(4), 153 (2015)
2. ATLAS Collaboration, Constraints on new phenomena via Higgs boson couplings and invisible decays with the ATLAS detector JHEP(2015, submitted)
3. G.C. Branco, P.M. Ferreira, L. Lavoura, M.N. Rebelo, M. Sher, J.P. Silva, Theory and phenomenology of two-Higgs-doublet models. Phys. Rep. **516**, 1–102 (2012)
4. A. Berlin, T. Lin, L.-T. Wang, Mono-Higgs detection of dark matter at the LHC. J. High Energy Phys. **06**, 078 (2014)
5. A. Azatov, S. Chang, N. Craig, J. Galloway, Higgs fits preference for suppressed down-type couplings: Implications for supersymmetry. Phys. Rev. **D86**, 075033 (2012)

6. ATLAS Collaboration, Search for neutral Higgs bosons of the minimal supersymmetric standard model in pp collisions at $\sqrt{s} = 8$ TeV with the ATLAS detector. J. High Energy Phys. **11**, 056 (2014)
7. Higgs to tau tau (MSSM), Technical Report CMS-PAS-HIG-13-021, CERN, Geneva (2013)
8. Search for the bb decay of the Standard Model Higgs boson in associated W/ZH production with the ATLAS detector. Technical Report ATLAS-CONF-2013-079, CERN, Geneva, Jul (2013)
9. E. Gross, O. Vitells, Trial factors for the look elsewhere effect in high energy physics. Eur. Phys. J. C **70**, 525–530 (2010)

Chapter 13
Conclusion

This thesis presents a novel analysis, searching for dark matter (DM) pair production in association with a Higgs boson which decays into a pair of bottom quarks. The analysis is performed using data from pp collisions collected at $\sqrt{s} = 8\,\text{TeV}$ with the ATLAS experiment at the Large Hadron Collider (LHC) at CERN, for an integrated luminosity of $20.3\,\text{fb}^{-1}$. Two techniques are employed, one in which the two b-quark jets from the Higgs boson decay are reconstructed separately (resolved), and the other in which they are found inside a single large-radius jet using boosted jet techniques (boosted). A set of increasing E_T^miss thresholds defines the final signal regions for each channel, optimized for individual signals in the parameter space probed. The resolved channel analysis consists of the focus of this thesis.

The results from the reserved channel are interpreted in the framework of a simplified model with a Z' gauge boson and two Higgs doublets, where the dark matter is coupled to the heavy pseudoscalar Higgs A, i.e., $Z' \to hA \to b\bar{b}\chi\chi$. In this model, Z' and A are produced on-shell, with electro-weak constraints imposed. An additional source of Higgs plus E_T^miss in this model comes from the decay of $Z' \to hZ$, $h \to b\bar{b}$, $Z \to \nu\bar{\nu}$.

By probing the resulting DM+Higgs($b\bar{b}$) final state requiring at least two b-jets in dedicated signal regions of 2 or 3-jets with large E_T^miss, we study a new kinematic regime different from existing dark matter searches at collider experiments. Both simulated and data-driven methods are used to describe the background processes, reaching good agreement of data and expected background across all control regions.

The data are found to be consistent with the Standard Model expectations in the signal region, with no excess of 2σ observed. Limits are set in the 2-D mass space of $m_{Z'}$–m_A, as well as $m_{Z'}$–$\tan\beta$ ($\equiv v_u/v_d$), for both the $Z' \to Ah$ dark matter production process alone, and the combined $h(\to b\bar{b}) + E_\text{T}^\text{miss}$ signature processes of $Z' \to Ah$ and $Z' \to Zh$. Results are given for using selections optimized for $Z' \to Ah$

alone, as well as using selections optimized for $Z' \to Ah$ plus $Z' \to Zh$ combined. The exclusion limits are strongest at low m_A and high $m_{Z'}$. The contribution from $Z' \to Zh$ allows for exclusions of regions with larger $\tan \beta$. In the $m_{Z'}$–m_A plane with $\tan \beta = 1$, $m_{Z'} = 700$–1300 GeV are excluded for m_A up to 350 GeV, with further exclusion of larger m_A for $m_{Z'}$ around 1200 GeV. In the $m_{Z'}$–$\tan \beta$ plane with $m_A = 300$ GeV, $m_{Z'} = 700$–1300 GeV are excluded for $\tan \beta$ up to 2, with further exclusion of larger $\tan \beta$ for $m_{Z'}$ between 800 and 1000 GeV. The limits are found to be in accordance with theoretical predictions of the LHC sensitivity. Upper limits on the visible cross-section for non-Standard-Model events in the signal regions for this analysis are also given. We expect to further probe this region with the prospects of more conclusive results with the 13 TeV run at the LHC.

CPSIA information can be obtained
at www.ICGtesting.com
Printed in the USA
LVHW02*1330110318
569449LV00002B/449/P